THE
SKY AT
NIGHT 10

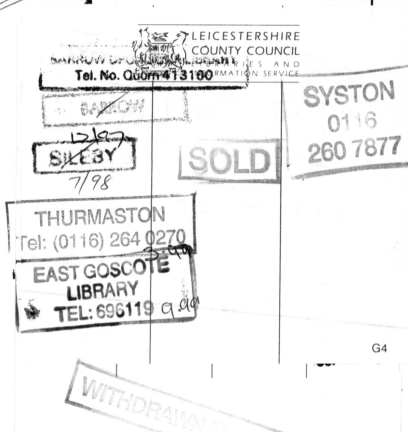

THE SKY AT NIGHT 10

Patrick Moore

JOHN WILEY & SONS
Chichester · New York · Brisbane · Toronto · Singapore

Wiley Editorial Offices

John Wiley & Sons Ltd,
Baffins Lane, Chichester,
West Sussex PO19 1UD, England

John Wiley & Sons, Inc.,
605 Third Avenue, New York,
NY 10158-0012, USA

Jacaranda Wiley Ltd,
G.P.O. Box 859, Brisbane, Queensland 4001, Australia

John Wiley & Sons (Canada) Ltd,
22 Worcester Road, Rexdale, Ontario M9W 1L1, Canada

John Wiley & Sons (SEA) Pte Ltd,
37 Jalan Pemimpin #05-04, Block B,
Union Building, Singapore 2057

Library of Congress Cataloging-in-Publication Data

Moore, Patrick.
   The sky at night / Patrick Moore.
       p.   cm.
   Includes index.
   ISBN 0 471 93763 0
   1. Astronomy—Popular works. I. Title.
   QB44.2.M664   1992
   520—dc20   92-36620
                              CIP

British Library Cataloguing in Publication Data

A catalogue record for this book is available from the British Library

ISBN 0 471 93763 0

Phototypeset by Dobbie Typesetting Limited, Tavistock, Devon
Printed and bound in Great Britain by Biddles Ltd, Guildford, Surrey

CONTENTS

FOREWORD

A pedestrian who found themselves walking along the Marylebone Road on a rainy night this April would have seen a rather curious assortment of people hurrying into the London Planetarium. Had he or she been a regular viewer of a certain monthly programme on BBC1 however, they would have immediately recognized the throng for what they were: the cream of British astronomy—professional and amateur alike—with the odd anonymous television producer bringing up the rear.

We had come together to celebrate the 35th anniversary of the first transmission of *The Sky At Night*. It was an occasion unprecedented in television history, the same programme presented by the same person continuously over three and a half decades. And there could hardly have been a more appropriate day for Patrick Moore to celebrate this achievement with his friends and colleagues. That morning's papers had been full of one of the most exciting astronomical discoveries of 1992: the ripples in the cosmic background radiation discovered by the satellite COBE. Stephen Hawking described the insight COBE gave scientists into the Big Bang as 'the discovery of the century'. Patrick and his guests at the Planetarium seemed hardly less enthusiastic.

You can read more about COBE and its discoveries in Chapter 35 (another coincidence?). As in previous *Sky At Night* books, Patrick has managed to combine wisdom and knowledge of interest to the beginner (Chapter 10, for example, has some practical advice on how to become a professional astronomer) with news of the very latest developments in the field. The freshness, clarity and topicality which have made his programme such an enduring success for us over the years, and which he and his producer Pieter Morpurgo still bring to each month's edition, are visible on every page.

It is a book which stands in its own right, and which can be recommended equally to devotees of *The Sky At Night* and to those who by reason of geography or age are unable to see the programme regularly. I discovered astronomy through Patrick's books long years before I was allowed to stay up late enough to watch him on television (I had to admit at our party for Patrick at the Planetarium that I only arrived on the planet a few days after the programme's third edition). I am certain that this book will give enormous pleasure, whether you intend to sit down and read it cover to cover, or dip into it chapter by chapter on evenings when clouds make more direct pursuit of astronomy impossible.

As for Patrick himself, let me thank him on behalf of BBC Television and his many loyal viewers and readers for being our cheerful and dependable guide to the night sky for so long. We look forward to the decades of *The Sky At Night* still to come, and wish him many more years of happy exploration of the heavens through the eye-piece of his telescope in Selsey.

November 1992   Mark Thompson,
                                            Head of Features, BBC Television

PREFACE

This is the tenth volume in the *Sky at Night* series. The pattern is essentially the same as before, so that I hope the text will take its readers through the period covered—autumn 1988 to late summer 1992. Though the chapters are based on the television programmes, they have been brought up to date, and in some cases very drastically amended in the light of later research. For example, one of our programmes (December 1991) dealt with the startling discovery of a new planet far beyond the Solar System; later it transpired that there had been a mistake, and that the planet does not exist at all!

Obviously I have combined many of the sections, and omitted some altogether. There is bound to be a certain degree of overlap here and there, because I have done my best to make each chapter self-contained, but I do not think that this overlap is very marked.

This has been a truly eventful time for astronomers, and I hope that you will join me in following it through.

Selsey, July 1992   PATRICK MOORE

ACKNOWLEDGEMENTS

My sincere thanks are due to all those who have joined me on *The Sky at Night* during the period covered by this book. In chronological order, they are: Dr Peter Cattermole; Paul Doherty; Dr Alan Wright; Dr Eleanor Helin; Dr Michael Disney; Dr Paul Murdin; Daniel Hofstadt; Dr Richard West; Professor Robert Wilson; Dr Ron Maddison; Douglas Arnold; Dr Nicholas Suntzeff; Dr Gabriel Martin; Dr William Kunkel; Professor Peter Willmore; Dr Ken Elliott; Dr Garry Hunt; Dr Edward Stone; Dr Torrance Johnson; Dr Ray Wilson; Harold Ridley; Dr Mike Hawkins; Dr David Carter; Professor John Baldwin; Professor Michael Rowan-Robinson; Dr Michael Penston; Dr Phil Charles; Dr Jasper Wall; Bruce Hardie; Professor Ulf Kussoffsky; Dr John Zarnecki; Professor Roger Bonnet; Dr Rüdeger Reinhard; Dr Robin Laurance; Dr David Dale; Dr Michael Perryman; Dr David Eaton; Dr Mart de Groot; Dr Brendan Byrne; Dr Jerry Doyle; Professor Ken Pounds; Professor Andrew Lyne; Professor Arnold Wolfendale; Dr Donald Yeomans; Dr Ian McHardy; Professor Alexander Boyarchuk; Dr John Mason; Dr Eric Chaisson; Dr Chris Blades; Professor Jim Westphal; Dr Terry Teays; Dr Duccio Macchetto; Dr Douglas Duncan; Dr Ron Parise; Dana Berry; Dr Ron Hanisch; Dr Stephen Maran; Dr David Crawford; Dr Derek McNally; Dr Matthew Bailes; Setnam Shemar; Dr Jerry Nelson; Dr Jerry Smith; Dr Malcolm Smith; Professor Iwan Williams; Dr David Malin; Professor Sir Bernard Lovell; Neil Armstrong; Commander Eugene Cernan; Dr Ellen Stofer; Dr Francis Jackson; Dr Manfred Grensemann; Dr Helen Walker; Professor Susan McKenna-Lawlor.

My thanks too to the publishers, for all their help and encouragement; and, needless to say, to all my many friends at the BBC, above all to Pieter Morpurgo—without whom there would be no *Sky at Night*.

PATRICK MOORE
Selsey, 28 July 1992

1 THE GEOLOGY OF MARS

Of all the planets in the Solar System, Mars is probably the most fascinating from our point of view. It is much less unlike the Earth than any of the others; and even though we could not live there in the open, there is every reason to believe that we will be able to send expeditions there in the foreseeable future. At the last General Assembly of the International Astronomical Union, held in 1991, the official Russian estimate, given by Dr V. Moroz, was 'some time between 2005 and 2010'.

On average Mars comes to opposition every 779 days. At opposition, the Sun, the Earth and Mars are roughly lined up, with the Earth in the mid position, so that Mars and the Sun are then opposite in the sky. Mars has a 'year' longer than ours, because it is further away from the Sun and is moving more slowly in a larger orbit; the period is 687 Earth days—equivalent to 669 Mars days or 'sols', because Mars' rotation period is somewhat longer than that of the Earth (to be precise, 24 hours 37 minutes 22½ seconds). Therefore, oppositions do not occur every year. Moreover, the Martian orbit is less circular than ours, and not all oppositions are equally favourable. They are best when Mars reaches opposition and perihelion simultaneously. The 1990 opposition was good, with a minimum distance from Earth of 47,800,000 miles—as against 53,000,000 miles for the opposition of 1993, while at the most unfavourable oppositions, such as that of February 1995, the minimum distance will be as much as 62,800,000 miles. And because Mars is small, with a diameter of just over 4000 miles, we have to make the best of its close approaches.

Telescopes show a reddish-ochre disk, with white polar caps and prominent dark markings. The reddish areas, which give Mars its naked-eye colour and led the ancients to name it after the God of War, are generally known as 'deserts'—which is not a bad name for them, even though they are totally unlike our deserts such as the Sahara. The dark regions are not old seabeds filled with vegetation, as used to be thought, but are simply areas where the reddish material has been blown away by Martian winds. The atmosphere is very tenuous—thinner than the Earth's air at a height of a hundred miles above sea-level; the ground pressure is below 10 millibars everywhere, and the atmosphere is made up chiefly of carbon dioxide. Temperatures are low. True, at the equator in midsummer they can rise to around 50 degrees Fahrenheit at noon,

Percival Lowell, at the eyepiece of the 24-in refractor at the Lowell Observatory, Flagstaff (Arizona).

Arroyo on Mars. A picture taken on 2 February 1972 from Mariner 9, from a range of 1033 miles. The arroyo—assuming that this is the correct interpretation—is 250 miles long, and 3 to 3½ miles wide. The centre of the picture is at Martian latitude 29° S, longitude 40° W.

but the nights anywhere on Mars are colder than we ever experience on Earth, partly because Mars is almost fifty million miles further from the Sun and partly because the thin atmosphere is very poor at blanketing in the Sun's warmth.

Unmanned missions have shown us that there are craters, mountains, valleys and giant volcanoes, one of which—Olympus Mons, or Mount Olympus—is three times as high as Everest, and is crowned by a 40-mile crater of the type known as a caldera. Two Viking probes made controlled landings on the planet in 1976, and searched for life, but found no sign of any, and it is now generally thought that Mars supports no life at all, at least at the present moment. The famous canals have, alas, been relegated to the realms of fiction; they were due to nothing more than tricks of the eye. (In my possession I have a cutting from the *New York Times*, dated 27 August 1911, with a glaring headline: MARTIANS BUILD TWO IMMENSE CANALS IN TWO YEARS. It would be exciting if it were true!)

But Mars will be a paradise for geologists, and so this may be the time to take a closer look at what we always term Martian geology, though technically it ought to be 'areography' (from Ares, the Greek equivalent of the war-god).

First, the warm hue of the deserts is deceptive; they are bitterly cold. There are two different types of terrain, each characteristic of one hemisphere. The southern two-thirds of the planet is heavily cratered,

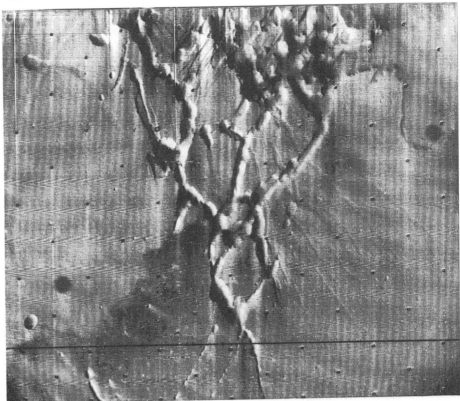

The 'Chandelier'—a huge valley system on Mars: the region is known as Noctis Labyrinthus. The area of the picture measures 336×264 miles. The picture was taken by Mariner 9 on 23 January 1972.

and is relatively high up; the northern one-third is smoother and lower. The zone which separates the two different regions is inclined to the Martian equator, but there are two exceptional upland regions, Tharsis in the western part of Mars and Elysium in the east. Tharsis intrudes into the cratered highlands of the west, and there are three huge volcanoes, Pavonis Mons, Arsia Mons and Ascræus Mons. Olympus Mons is also in the Tharsis area, though some distance away from the main ridge. In the south there is another famous feature, the deep basin Hellas, which is easy to see with a small telescope when Mars is well placed, and which was once thought to be a lofty, snow-covered plateau.

Why are the two areas so different? It seems that the cratered hemisphere is very old, perhaps with an age of around 4000 million years, and that early in Mars' history there was some event which destroyed the craters in the north, resurfacing huge tracts and producing the smoother plains we now see. Lofty volcanoes rose, and since these are confined mainly to Tharsis and Elysium they were presumably associated in some way with the uplifting of these areas.

Tharsis is dominant. It is about the same size as Africa south of the Congo River, and has a tremendous variety of geological features. First, there are the spectacular shield volcanoes, similar to those of our Hawaii but much larger and more massive. Thousands of long lava-flows radiate out from their summits, as well as fractures which must have opened as the Martian crust was stretched during the formation of the Tharsis bulge. Here too is the amazing Valles Marineris, a system of canyons which runs along the equator for a distance equal to that between Dublin and Cairo. Some of the canyons are as much as 5 miles deep and 125 miles wide, so that they dwarf our much-vaunted Arizona Canyon. On the flanks of Tharsis we find the network of Noctis Labyrinthus, from which the principal canyons emerge before crossing the central desert. Landslides on some of their floors are as large as a British county such as Kent.

There is no firm evidence that water ever flowed across the floor of Valles Marineris. Probably the fracturing of the Martian crust triggered off canyon formation, after which there was a slow wasting-away of the rocks. On the other hand there is conclusive evidence elsewhere of past water activity; we can see obvious dry riverbeds and even islands, where the flow of currents around resistant objects has left small tear-drop islands behind. Moreover, some of the channels have dendritic tributaries, rather as we find on Earth. Because most of the channels occur in the heavily cratered areas, they must be ancient. Thousands of small channels are found between the southern craters, but the biggest systems are those around the Chryse Basin (where the lander of Viking 1 came down) and near the zone which divides the two types of terrain. There are also outflow channels, such as Juventæ Chasma, which cuts a sixty-mile wide swath across the smooth Lunæ Planum. Juventæ, and other features like it, appear to have been cut by catastrophic floods, due perhaps to the melting of sub-surface ice by volcanic outbreaks.

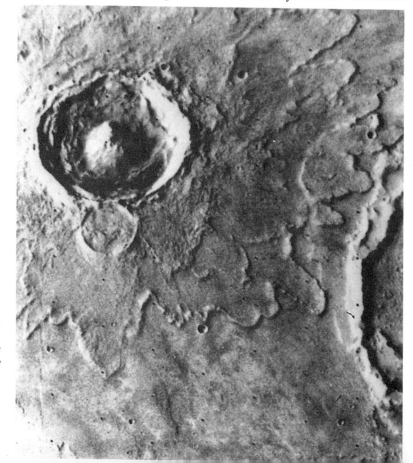

The Martian crater Yuty, 11 miles in diameter; latitude 22°4' N, longitude 034°1' W. This is some way from Chryse, where the Viking 1 lander came down. This picture was taken by the Viking 1 orbiter (NASA).

Yet there can be no liquid water on Mars now, because the atmospheric pressure is too low. Therefore, the volatiles must be imprisoned below the surface in the form of ice. Theory predicts that Mars must have outgassed considerable volumes of water early in its history, and that much of this did not escape into space, so that we may find ice at a depth of a few inches within 40 degrees of the poles. It follows, therefore, that the Martian climate has changed with time, and that the sub-surface is full of pore spaces into which the volatiles have seeped, to be frozen there and preserved.

The polar caps are of vital importance in all studies of Mars. Curiously, they are not identical. During Martian spring and summer they shrink, and this means the depletion of the upper ice, which is carbon dioxide ice rather than water ice. When this happens in the northern hemisphere, the atmosphere is usually clear, and the temperature rises high enough for all the carbon dioxide to sublime (i.e. change directly from the solid to the gaseous state), exposing a residual cap of water ice beneath. In the southern hemisphere the atmosphere is 'dustier' during the period when the cap is shrinking, so that the water ice residual cap—assuming that it exists—is never seen. Around both caps we find large areas of layered deposits which may be up to three miles thick, and appear as massive scarps emerging from beneath the carbon dioxide ice. Close-up probe pictures show that they are made up of light and dark layers, presumably hundreds of feet thick, which are probably composed of dust and volatiles which will record fluctuations in the Martian climate over very long periods.

The main geological difference between Mars and the Earth is that on Mars the main activity ended long ago; Mars was so small that it was unable to stay dynamic, as the Earth has done, and it could not retain a dense atmosphere. Neither is there any Martian equivalent of terrestrial plate tectonics. On Earth, the uppermost surface layer (the lithosphere) is moving slowly above the underlying mantle. Volcanoes such as those of Hawaii slip away from their 'hot spot' in the mantle below, so that they cease to erupt—as Mauna Kea in Hawaii has done. On Mars this does not happen, so that the volcanoes stayed above their 'hot spots' for immense periods, and were able to grow to tremendous sizes.

There does not seem to be much chance of active vulcanism on Mars now, but we cannot be sure. We need a 'Rover' capable of sampling and analysing materials from various sites; we need radar surveys, so that we can, for example, locate the presence of ice (which blocks out a radar beam). Above all, we need an automatic sample and return probe, so that we can analyse Martian materials in detail and find out, once and for all, whether there has ever been life there.

All these may come—perhaps sooner than many people expect. Meanwhile, Mars remains an intriguing world; we are learning more all the time, and it is quite likely that the first man on Mars has already been born.

2 THE NOVICE OBSERVER

The run-off shed covering my 12½-in reflector at Selsey, in Sussex. The shed is in two halves, which can be run back in opposite directions.

Over the years, I have had many hundreds of letters asking for my advice on how to begin observational astronomy, and what equipment is needed. In *The Sky at Night* we have presented plenty of programmes about telescopes, and I have always stressed that in my view no telescope is of much use astronomically unless it has an aperture of at least 3 inches (for a refractor) or 6 inches (for a Newtonian reflector). Of course smaller telescopes can be bought, often for a few tens of pounds, but they are always disappointing; they have restricted fields of view, low light-grasp, and (nearly always) rickety mountings. Rather than purchase a very small telescope, I would unhesitatingly prefer good binoculars.

Larger telescopes need observatories. My 12½-inch reflector has a run-off shed, which has been in action ever since 1947 and is very satisfactory; my 15-inch reflector has a 'dome' which looks rather like an oil-drum, while my 5-inch refractor has a shed with a run-off roof. But many people have asked me the wisdom, or otherwise, of erecting an observatory on the flat roof of a house, and again I always give the same answer: *Don't.*

The reason is quite straightforward (apart from the general instability: it would be only too easy to walk off the edge of the roof at the dead of night). A dwelling-house is presumably warmer than its surroundings at night. Heated air will swirl up, and ruin the seeing completely. If you want to check this, all you have to do is to observe a lamp on the far side of an electric fire. The air rising from the fire will make the light flicker wildly—and the same will happen to starlight when observed from a rooftop observatory.

Neither is it profitable to observe by poking a telescope through a bedroom window. Again heated air is a nuisance, and there are also other hazards; I once dropped an eyepiece forty feet on to a concrete path (though in my defence, I must add that I was only eight years old at the time).

Dark adaptation has also to be borne in mind. Recently I was working in my study, typing on my ancient Woodstock (1908 vintage) until the early hours, after which I decided to see whether the sky had cleared. It had, but for several minutes I could see no stars at all. Generally, one has to 'dark adapt' for at least half an hour before making the best use of a good night, though the time needed is bound to vary with different people.

My 5-in refractor at Selsey, The 'observatory' has a run-off roof.

There is the extra problem of recording observations, which cannot be done in pitch darkness. Use a red light, which is less damaging than a white one. It is helpful to fix a red light to a clipboard, and there are also devices which are really pen-torches but which are fitted with pencils. I find mine invaluable. If you are not sketching, but merely making notes, the obvious solution is to use a portable tape recorder.

Comfort is essential. The trouble is that one sometimes has to observe under awkward conditions—for example, when timing an occultation; the Moon may be inconveniently low, or else obscured by a tree, in which case the only remedy is to use a portable telescope. Hazards can be encountered here, too. I once made what I thought was a good timing, but as soon as I had finished I stepped back, fell over the cat, and closed my hand convulsively on my stop-watch, re-starting it. Always make sure that you know where everything is—even the cat—because if you do not, then you will have to use a light to grub around, and your dark adaptation will suffer.

A small refractor can be awkward because when the target object is high, a telescope on a tripod will have its eyepiece low down, and the observer will have to be something of a contortionist. Elbow eyepieces may help, but personally I do not like them; every time a ray of light passes through a piece of glass, is reflected from a mirror, or reflected from a prism it is slightly enfeebled. I would prefer to mount the tripod on a box or something of the sort, provided that it is firm. With a Newtonian this problem does not arise, but unless a long-focus Newtonian has a revolving head (as my 15-inch has) the eyepiece will be high up at times, as well as being at an awkward angle, so that steps or a ladder must be used. I prefer massive observing steps; there is less danger of falling off them.

Now for actual observing. One golden rule is: Never send away your originals. I once did so, and a stack of my best planetary drawings is still presumably wandering around in the bowels of the Post Office—and this was years ago, when the post was much more reliable than it is today. Always keep your originals, and put them in a file or book. I keep a separate observing book for each subject in my regular programme: one for the Moon, one for Mars, one for variable stars and so on.

Some observers make drawings of the Moon and planets directly at the telescope. I am not skilful enough for this, so I make my drawing, come indoors and put the 'fair' copy in my book, and then return to the telescope to check. But remember—always enter your observations immediately. Resist the temptation to 'leave it until tomorrow', because mistakes are bound to creep in.

Of course, there are times when an observation has to be hurried, usually because of imminent cloud. In this case make the most important observations first, and trust to luck that you will have enough time to finish off and re-check.

As a planetary observer, I use prepared blanks. Jupiter, for example, is obviously flattened. You can draw a correct shape each time; but why do so, when a prepared blank is much easier and probably much

Blank for observing Jupiter. If you want to make a drawing of the planet, I suggest that you photocopy this flattened disk and use it as an outline.

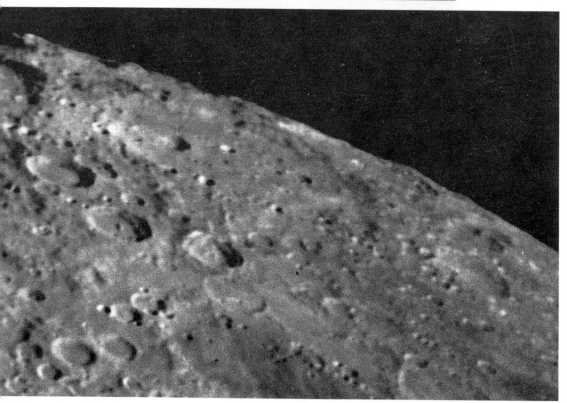

The lunar crater Bailly; it is 185 miles across, but lies near the Moon's limb as seen from the Earth, and is very fore-shortened. Photograph by Commander H. R. Hatfield, with his 12-in reflector.

more accurate? The phases of Mars are also predictable; they range from 100 per cent at opposition down to only 85 per cent at quadrature, and this must always be taken into account if you want to make your drawings really accurate. With Venus and Mercury, admittedly, things are less easy, and Saturn is a real problem; the better the quality of artistry, the better the results—as I know to my cost!

Avoid prejudice. It is only too easy to 'see' what you half expect to see (remember the canals of Mars). Also, do not be too quick to reject an observation. I once thought I had glimpsed the two satellites of Mars, Phobos and Deimos, with my 12½-inch reflector, but when I worked out the actual positions they were wrong. Yet I could 'see' them, and I noted them as 'recorded but presumably erroneous'. Actually the error lay in my calculations; I had added up 2 and 2 and made the answer 2. The observation itself was quite correct.

Do not use too high a magnification. Remember that a smaller, crisp image is far better than a larger, blurred one. As soon as there is any loss of definition, change to a lower power.

All this applies to planets. With variable stars, of course, no drawing is involved, and it is a matter of comparing the variable with nearby comparison stars. I invariably use a tape recorder, which means that I do not have to use a light at all—and with stars at the limit of the telescope's power, dark adaptation is all-important.

With the Sun, the only sensible course is to project the image; I wonder how many times I have stressed the danger of looking direct with any optical appliance, even with the addition of a dark filter? The Moon is harmless, but there are pitfalls here too. Trying to draw a large area of the lunar surface at any one time is hopeless, and the best thing to do is to concentrate upon a small region, making the sketch really precise. When I began to take a real interest in the Moon, I obtained an outline map and spent over a year making at least three sketches of every named formation, since the changing angle of sunlight means that a crater or peak can alter dramatically in appearance even over a period of a few hours. The actual drawings were useless, of course, but at the end of the programme I knew my way around the Moon!

Finally, there are some essential data to be added to each observation: date, time in GMT, type of telescope, magnification, name of observer, and seeing conditions, usually on the Antoniadi scale (from 1, near-perfect, down to 5, so poor that observations would not be attempted except for some special phenomenon which will not recur). If any of these data are omitted, the observation promptly loses most or all of its value.

These notes are aimed at the absolute novice. By the time you have reached the stage of serious photography, or using photometric or electronic equipment, you will long since have learned what to do and what not to do. Yet everyone has to make a start, and at least I have one qualification: I have made almost every mistake that it is possible to make, so that I believe I can help others to avoid falling into the same absurd traps.

3 RADIO STARS

Radio astronomy is a young science, but during the past few decades it has become of supreme importance. In this, Britain has led the way; who does not know about the 250-foot 'dish' at Jodrell Bank, now renamed in honour of Sir Bernard Lovell, who master-minded it? Today, the Lovell Telescope is the focal point of a whole network, MERLIN, made up of several individual telescopes at Knockin, Cambridge and elsewhere.

Ideally, radio astronomers would like really large telescopes, covering much of England. Since this is rather obviously out of the question, there must be a compromise. Instead of building a 'Great British Dish', sections of it are constructed—each section being in the form of an individual telescope. In principle, the largest parts of each circular band are left out, and use is made of the fact that the Earth rotates, causing each telescope to sweep out a circle across the sky (or, to be more precise, an ellipse—unless your telescope happens to be centred upon the North or South Pole). Provided that the energy collected by the separate telescopes can be brought together, they will act almost as well as the complete Great British Dish would have done.

Consider, for a moment, just two telescopes. Radio waves arriving from any particular direction will usually get to one of the dishes before the other. The difference in arrival time depends upon the position of the radio source, and if these times are monitored accurately enough the position and size of the source itself can be found.

Making the various telescopes work together is a major problem. They can either be connected by a radio link, as with MERLIN, or we can use video-type recorders to collect the separate signals, after which the tapes are brought together and played back. The problem here is that we need very accurate time-signals to synchronize the tapes. Radio astronomers regularly use hydrogen maser clocks which are so accurate that they could run for millions of years without gaining or losing more than the tiniest fraction of a second.

The new Australia Telescope, inaugurated in 1988, uses a combination of both these methods. The group of six dishes at Culgoora in the northern part of New South Wales gives the gross structure of the radio sources, and will be connected by fibre-optic links. The dishes further out—such as Mopra, near Siding Spring, and Parkes near Coonabarabran—will use recorders. Other radio telescopes can also be brought into the array, such as the dishes near Hobart in Tasmania and the NASA dish near Canberra.

Sir Bernard Lovell at Jodrell Bank, 1990.

Next, what exactly are these cosmic radio sources?

Initially it was assumed that the main sources would be bright stars such as Sirius, Rigel and Vega, but this was soon found not to be so; brilliant stars were not powerful sources. Instead, the radio waves came from objects which were optically dim—supernova remnants (such as the Crab Nebula) and certain types of galaxies, as well as quasars, now known to be the nuclei of very remote galaxies which could well be powered by central black holes. In fact, radio telescopes are best at locating objects which are either very big or else very hot, with temperatures of millions or thousands of millions of degrees. Stars are essentially point sources, and if our Sun were placed at, say, half a dozen light-years, it would be undetectable by our present radio techniques. The old term 'radio stars' for most sources has long since been dropped.

However, it has now been found that there are certain types of stars which do show up at radio wavelengths. In particular there are the P Cygni stars, of which the prototype (P Cygni itself) is easily visible with the naked eye near the centre of the Cross of Cygnus. It is very remote, and also very massive; moreover it seems to be doing its best to blow itself to pieces.

All stars produce large amounts of energy near their centres, and this energy has to fight its way through to the surface. The largest stars produce very much more energy than less massive ones; a star twice as massive as the Sun will produce eight times the amount of energy. P Cygni, more massive still, is emitting so much energy that it is literally blowing off its outer layers at a rate of over a thousand trillion tons

The 210-foot radio telescope at Parkes, New South Wales, as I photographed it in 1987.

of material per second. Radio maps have shown this expelled gas at large distances from the star, and have allowed us to build up a good picture of what is going on.

Next there are the 'spotty' stars, of which the prototype is RS Canum Venaticorum in the little constellation of the Hunting Dogs. Here we have a binary system, made up of a fairly normal solar-type star together with a larger red companion. Because they are so close together, the gravitational effect of the yellow star stirs up the interior of its red companion, causing large spots to appear on its surface—not unlike our own sunspots, but on a much grander scale. These starspots consist of a violent mixture of strong magnetic fields and fast-moving atomic particles, producing bursts of radio energy with temperatures measured in thousands of millions of degrees.

Amateur astronomers have been of great help here. In Australia, the group led by Colonel Arthur Page at Mount Tamborine Observatory, Queensland, has produced optical observations of the RS Canum Venaticorum stars which fit in perfectly with the radio data. As the binary system rotates, so does the darker starspot area. Although the starspots are too small to be seen directly (after all, the star itself looks only like a speck of light), the magnitude drops when the darker, spotted area is facing us.

Then there are the fascinating 'symbiotic stars', which are different again. A typical example has recently been studied from Australia by Alan Wright and David Allen, who have come up with some startling conclusions.

There seems to be a dirty, dusty, red giant star which is blowing away its outer layers. Lurking nearby sits a tiny white dwarf star which has used up all its nuclear energy. The dwarf becomes covered with this dusty 'wind'; the captured gas provides a new source of energy for the worn-out white dwarf, and gives it a new lease of life. This renewed activity makes the white dwarf eject large amounts of its own material in what is called a helium flash. An enormous battle then rages between the 'dirty red giant' and the 'flashing white dwarf'. First the one, then the other gains the upper hand, but eventually the red giant wins, and the white dwarf is destroyed. Subsequently the red giant will itself enter the white dwarf stage, but by then it will no longer have a companion. Yet another battle in the Star Wars saga will have been completed . . .

Obviously we are learning more all the time. Just as we have come to the stage of building new types of optical telescopes, so we are also producing new generations of radio telescopes, and it should not be long before we can even make use of 'outstations' in space—eventually on the surface of the Moon. It is a staggering thought.

4 MISSILES FROM SPACE

M ost people have heard of the asteroids, those tiny worlds which have been described as 'planets in miniature'. The main swarm lies between the orbits of Mars and Jupiter, but there are some which venture closer-in to the Sun, and these mavericks have been the subject of considerable discussion recently.

It is usually said that the first asteroid known to come within the orbit of Mars was No. 433 Eros, discovered by Witt, from Berlin, in 1898. This is not quite true—132 Æthra, found by Watson in 1873, can come just within the path of Mars when it is at perihelion—but Eros swings much closer-in, and can pass the Earth at only about 15,000,000 miles, as it did in 1931 and again in 1975. At the 1931 return it was carefully monitored, because there is a mathematical method of using it to measure the distance between the Earth and the Sun—though the method is now obsolete, and in this respect at least Eros has lost its importance. It is sausage-shaped, perhaps 25 miles long but only 9 miles wide.

Many asteroids have now been found which can approach us much more closely than Eros. The leading hunter of these cosmic missiles is Dr Eleanor Helin, of the United States, who began her astronomical career as a planetary geologist and gravitated to the studies of meteorites and small asteroids. Her work has confirmed that these tiny asteroids are much commoner than has been believed. Even the largest known, 1036 Ganymed, has a diameter of less than 25 miles.

Basically, there are three types of close-approach asteroids: the Amors, Apollos and Atens, all named after the first-discovered member of their class.

An Amor asteroid has its perihelion outside the orbit of the Earth, so that it never actually crosses the Earth's orbit, though it may go out as far as the path of Jupiter. Amor itself was discovered in 1932, and is a real midget just over half a mile across. In fact another Amor asteroid, 887 Alinda, had been discovered earlier (in 1918) but had been lost, and was not recovered until 1969. There was also 719 Albert, discovered in 1911. It was no more than 1½ miles in diameter, and, sadly, it has been lost; whether it will ever be seen again is questionable.

An asteroid is given a number only after it has been under observation for long enough to have its orbit well worked out. Albert has the unhappy distinction of being the only numbered asteroid to have 'gone missing'

Comet IRAS–Araki–Alcock, 1983. The camera was following the comet, so that the stars are drawn out into trails. The comet was discovered independently from the IRAS satellite and by two ground-based observers, George Alcock in England and Araki in Japan.

probably permanently. 878 Mildred, found in 1916, was almost given up, but was found again in 1990 after a long hunt, so we must not abandon Albert, but we have to admit that the outlook is not at all promising.

Apollo asteroids do cross the Earth's path, but have orbits which are larger than ours, so that for most of the time they are further away from the Sun than we are. Apollo itself (diameter nine-tenths of a mile) was discovered in 1932, and then lost until 1973, but it has since been so carefully observed that it is unlikely to be mislaid again.

Much the largest of the Apollos is 2212 Hephaistos, with a diameter of 5½ miles. Hephaistos has been singled out by two eminent British astronomers, Victor Clube and Bill Napier, as their 'Cosmic Serpent' of ancient times. Of the rest, very few are as much as two miles across.

Some of the Apollos have very eccentric orbits, and two of them, 1566 Icarus and 3200 Phæthon, invade the torrid regions within the orbit of Mercury. Icarus can go within 17½ million miles of the Sun, and Phæthon ventures even closer, at 13 million miles. At perihelion they must be red-hot. At aphelion (greatest distance from the Sun) Icarus recedes to 183,000,000 miles, well beyond Mars, and will then be bitterly cold, so that it must have about the most uncomfortable climate in the Solar System. Phæthon is less extreme, and never goes out to more than 102,000,000 miles from the Sun. Neither has a diameter of as much as a mile.

Finally we have the Aten asteroids, of which less than a dozen are known and of which only six have been sufficiently well-observed to receive numbers: 2062 Aten, 2100 Ra-Shalom, 2340 Hathor, 3362

Khufu, 3554 Amun and 3763. Ra-Shalom may be as much as a couple of miles across, but Aten is a mere half-mile, and Hathor has the distinction of being the smallest object to have received a name; its diameter is about 350 yards!

What are the main points of interest about these tiny bodies?

For one thing, they provide opportunities for possible rocket mission rendezvous. Of special interest is the 380-yard asteroid 3757, which has actually been contacted by radar from the Arecibo 'dish' in Puerto Rico. Eleanor Helin and her colleagues have found that there are several chances for rocket missions between now and the year 2010, particularly in 1995, 1997 and 2007, so that something of the sort may be attempted. Not that Asteroid 3757 is the only candidate; there are various others.

Aten itself was one of Eleanor Helin's discoveries, on 7 January 1976, when it was just over 11,000,000 miles away. Like all modern discoveries, it was found photographically, and it is always too dim to be seen with average amateur-owned telescopes.

If these small asteroids can come so close to us, is there any danger of a collision? The answer must be 'yes', and, for that matter, it may have happened in near-recorded times. More than 22,000 years ago an object hit the Arizonan desert and blasted out the now-famous Meteor Crater. It may be said that the Meteor Crater missile could have been ranked as a small asteroid, since its size was probably much the same as that of Hathor.

Tiny asteroids are susceptible to all manner of perturbations, and in some cases their orbits are unstable. Such is the case with 1675 Toro,

Fragment of the Barwell Meteorite, which landed in Leicestershire on Christmas Eve, 1966.

a dwarf measuring 2.9 by 2.4 miles, discovered in 1948. It has an orbital period of 1.6 years, which is exactly 8/5 of that of the Earth and 13/5 that of Venus. The distance from the Sun ranges between 71,700,000 miles and 182,300,000 miles, so that it crosses the orbits of both Earth and Mars and can pass near Venus. This will not be the case indefinitely. Even if it does not hit either Earth or Venus, a single close encounter with Mars will alter its orbit completely in about three million years.

It seems inevitable that sooner or later the Earth will again be struck by an asteroid, and the results could be catastrophic. There have been well-supported suggestions that such a hit, about 65,000,000 years ago, caused such widespread devastation that the dinosaurs, which had ruled the world for so long, were unable to cope with the new climatic conditions, and died out. Whether this is true or not remains uncertain (personally, I am somewhat sceptical), but it is at least a distinct possibility, and it is significant that deposits laid down at about this time contain more iridium than would normally be expected—and iridium is a characteristic of some of the Solar System's junior members. Sir Fred Hoyle has even suggested that we ought to take precautions against a cold period caused by excessive cloudiness following an impact, by building huge machines to stir up the Earth's oceans and keep them at a stable temperature until the crisis is over; there has also been a scientific study to see whether an approaching asteroid could be destroyed or diverted by a nuclear bomb. Whether such drastic measures will be needed in the foreseeable future remains to be seen!

The asteroids: Gordon Taylor, Professor Alec Boksenberg, and Dr Eleanor Helin all have asteroids named after them.

It has also been suggested that close-approach asteroids may simply be the nuclei of dead comets, which have lost all their volatiles. It is true that Phæthon moves in much the same orbit as the Geminid meteor stream, and could well be the 'parent' of this stream. Asteroid 2201, Oljato, has an exceptional spectrum and has been suspected of affecting the interplanetary magnetic field, so that it could have a tail of tenuous material millions of miles long; another candidate is 2101 Adonis. Support for this idea came in 1992, when it was found that Asteroid No. 4015 was identical with an object observed in 1949 and then classed as a comet: Wilson–Harrington. Unquestionably it had a short tail in 1949 which it has now lost. The very eccentric-orbit asteroids such as 944 Hidalgo, which moves out almost as far as Saturn, may also be ex-comets, though the spectra do not seem to fit in well with such an idea and no trace of any coma or tail has been seen.

Certainly the tiny Earth-approach asteroids are of real interest. They are surprisingly numerous, and more of them are being found every year.

5 SOME MINOR MYSTERIES

Centuries ago, it was widely assumed that the heavens are unchange-able. Nothing could ever be further from the truth. There are changes going on all the time, both in the long and in the short term. And this leads on to one or two 'mysteries' which are admittedly of minor importance, but which are interesting none the less.

One of these relates to Sirius, the Dog Star, which is by far the most brilliant star in the sky—half a magnitude brighter than its nearest rival, the southern Canopus. Not that Sirius is a true cosmical searchlight; it is 'only' 26 times as luminous as the Sun, whereas according to one recent estimate Canopus could match 200,000 Suns. But Sirius is a mere 8.6 light-years away, and is one of our nearest stellar neighbours. It has a White Dwarf companion, with only 1/10,000 the luminosity of Sirius itself. Not unnaturally, Sirius B is often nicknamed 'the Pup', but it is a very massive pup; though its diameter is no more than 26,000 miles, less than that of a planet such as Uranus or Neptune, its mass is equal to that of the Sun. Its density is at least 60,000 times that of water.

Sirius has an A-type spectrum, and is pure white. Yet in many old records it is described as a red star. Why is this? Has there been some definite change, or are the old records wrong?

The last great astronomer of ancient times was Ptolemy of Alexandria, who flourished between AD 120 and 180. Of his life and personality we know nothing, but he was most certainly an expert theorist and observer, and periodical attempts to discredit him have been signally unsuccessful. His great work, the *Almagest*, has come down to us by way of its Arab translation (which is why we use its Arabic name), and is really a compendium of ancient science; without it, we would know much less about Classical research than we actually do.

In 1760 a British writer, Thomas Barker, drew attention to the fact that Ptolemy had listed six stars which were very definitely red. They were Arcturus, Aldebaran, Pollux, Antares, Betelgeux, and—Sirius. Of these, five are definitely reddish or at least orange, but this cannot be said of Sirius, and Barker concluded that there had been a change. He also cited Seneca, who lived from 4 BC to AD 65: 'The redness of the Dog Star is more burning; that of Mars is milder'. Later, the strange and unpopular American astronomer T.J.J. See undertook some research, and found that Sirius had also been described as red by various other ancient observers ranging from Homer to Pliny.

Sirius: the spikes are photographic effects, but the tiny White Dwarf companion is shown. The Companion is only 1/10,000 as bright as Sirius itself.

It is, of course, true that Sirius twinkles violently, because of its brilliance; this is particularly evident from Britain, where it is never high in the sky. But years ago I carried out an experiment in which I asked viewers of the *Sky at Night* to look at Sirius and estimate its colour. Virtually nobody described it as red. And from countries such as Egypt, where Ptolemy lived, Sirius rises higher and therefore twinkles less.

What can be the explanation? Sir John Herschel, in 1839, suggested that there might have been an effect due to interstellar matter between Sirius and ourselves, which has now moved away; but this does not seem likely, and Sirius itself is a normal Main Sequence star which would not be expected to change much over a period of only a couple of thousands of years.

On the other hand, we do have the White Dwarf companion, and we know that before a star becomes a White Dwarf it goes through a Red Giant stage. Inevitably, it has been suggested that in Ptolemy's time the Companion was of this type. Yet the present Sirius, combined with a Red Giant companion, would make a 'star' almost as bright as Venus, and there is no hint of this in the old writings. Moreover, the time-scale is all wrong—and for the same reason, it seems highly improbable that the Pup went through a temporary Red Giant stage following some sort of outburst.

To me, the final word comes from Chinese observations made by Sima Qian in the first century BC. Qian calls Sirius 'white'. Therefore, the balance of evidence seems to be strongly against any real alteration. It is true that in ancient times Sirius was regarded as an unlucky star, and there has been a recent suggestion that this was the reason for its being described as red: red for danger! But surely this is straining the evidence too far.

Another of Ptolemy's 'red' stars was Pollux, in Gemini, which has a K-type spectrum and is today no more than orange. It was once ranked as inferior to its neighbour Castor, and as recently as 1700 the first Astronomer Royal, John Flamsteed, rated Castor as of magnitude 1 and Pollux only 2, but apparently Flamsteed made only one observation. Castor is a multiple system, whereas Pollux is single, and the two are not genuinely close together; Castor is some 10 light-years the more remote. Again the evidence for change seems very slender indeed.

However, let me now turn to a puzzle which is much less easy to solve. It concerns Mizar or Zeta Ursæ Majoris, the second star in the 'tail' of the Great Bear.

It is easy to see that there is a much fainter star, Alcor, close to Mizar. The distance between the two is 709 seconds of arc; the magnitudes are 2.3 and 4.0, and Alcor is visible without optical aid on any reasonably clear night. It and Mizar are often nicknamed 'Jack and his Rider'. Yet the Arabs of a thousand years ago regarded Alcor as a very difficult naked-eye object. In the thirteenth century Al-Kazwini stated that 'people tested their eyesight by this star', and in the fourteenth century Al-Firuzabini called it 'The Test'.

This is certainly peculiar. The Arabs were notoriously keen-sighted, and their skies were much less light-polluted than ours. So what is the answer? Mizar itself is a well-known binary (each component of which is itself a spectroscopic binary) but neither it nor Alcor seems to be the sort of system to vary appreciably over a period of a few centuries.

It has been claimed that the Arabs were referring not to Alcor, but to an 8th-magnitude star which lies between it and the Mizar pair. According to W.H. Smyth in his famous book *Cycle of Celestial Objects*, this star was first noted by D. Einmart of Nürnberg in 1691, and was reobserved in 1723 by a German astronomer who named it 'Sidus Ludovicianum' in honour of the Landgrave of Hesse—believing it to be either a new star or even a planet (!). It is not genuinely associated with the Mizar system, and lies in the background; it is well below naked-eye or even binocular visibility. It is unlikely to have brightened up since Arab times, but it is equally unlikely that the Arabs were wrong. This is certainly a minor mystery which I do not pretend to be able to solve.

Incidentally, Megrez or Delta Ursæ Majoris is much the faintest of the seven stars in the Plough pattern; it is of magnitude 3.3, well below its six companions. Yet the great sixteenth-century Danish astronomer Tycho Brahe ranked it of magnitude 2. It may be slightly variable, but there is no proof, and certainly no hard and fast evidence of any secular change.

Finally, let me consider Messier 42, the Great Nebula in Orion, which is the most famous of all the gaseous nebulæ, and is an easy naked-eye object; it is a stellar birthplace, and fresh stars are being created inside it. (It also contains some strong infra-red sources, which are now believed to be immensely powerful stars whose 'visible' light can never reach us because of the intervening material.) M42 was first described in 1610

The Great Nebula in Orion, photographed with the Palomar 200-in reflector.

by an otherwise obscure astronomer named Nicholas Peiresc. This was shortly after the invention of the telescope. Neither the Greeks, the Arabs, nor even Tycho Brahe mentioned it at all—and yet how could they have overlooked it? I refuse to believe that it increased in brightness at the precise moment when the telescope was developed, but I also fail to understand how it can have escaped the eagle eyes of Tycho and his predecessors. Again we have a minor problem which is not easy to solve.

These are only a few of the many lesser-known mysteries of the stellar sky. Our knowledge may not be nearly so complete as we like to believe.

6 HITCH-HIKER

In astronomy, as in so many other subjects, things are not always what they seem. To give just one example: in Victorian times it was generally thought that most of the stars were more luminous than our Sun, perhaps by a factor of thousands. Of course, there are many stars which far outshine the Sun—Rigel, for instance, by at least 60,000 times—but we have now found that most stars are actually much less powerful than the Sun. The Victorians were misled by what is termed 'observational selection'. They saw the bright stars, which were the more obvious, and did not appreciate the large numbers of faint ones.

It now appears to be much the same with galaxies, and the work carried out at Cardiff University by Michael Disney and his colleagues indicates that observational selection is very marked. Moreover, their investigations have uncovered large numbers of galaxies of entirely new type, not identified earlier because of their low surface brightness.

There are three ways in which galaxies may be overlooked:

1. They may simply be too faint.
2. They may be confused with stars.
3. Their surface brightness may be lower than that of the general sky background—remembering that extragalactic sources make up only about 1 per cent of the overall luminosity of the night sky. So we are rather in the situation of someone in a brightly lit room peering out through a window at the darkened landscape. Try it, and you will find that you will not see much—apart from any other bright lights.

The surface brightness of a galaxy determines its ease of observability. If it is too bright and too condensed, it will be confused with a star; if it is too faint or diffuse, it will be missed altogether. At Cardiff, Disney and his colleagues have drawn up a graph which shows this—and they have found that the numbers of galaxies actually observed fits in almost exactly with the graph showing ease of observability! Therefore it seems that many systems must have been missed, and we realize now that this is indeed so.

Much of the work has been carried out with the Automatic Plate-Measuring Machine devised at Cambridge by E. Kibblewhite. What this does is to scan a photographic plate with a laser beam, and then store the results in a computer for analysis. After a good deal of what may be termed 'electronic massage', a star can be distinguished from a galaxy,

The Whirlpool Galaxy, M51 in Canes Venatici (the Hunting Dogs), photographed with the Palomar 200 in reflector. This is a beautiful spiral, and was the first spiral to be described (by Lord Rosse in 1845, using his 72-in reflector at Birr Castle in Ireland).

and very low surface brightness galaxies (LSBGs) can be brought out. The Cardiff workers have paid particular attention to one of the nearby clusters of galaxies, in the constellation of Fornax. On one plate, showing many ordinary galaxies, they found no less than 1700 new ones which had not been previously identified. Either they had been mistaken for stars, or else they had been completely overlooked.

It has also been established that by no means all the LSB galaxies are dwarfs. Far from it! For example, what seemed to be a very small, dim system was photographed by David Malin at the Siding Spring observatory in Australia. Elaborate photographic techniques showed that it is in fact a vast system, covering a hundred times the area of our Galaxy and containing ten times as many stars—but it is so spread-out that from its distance of around 2,500,000,000 light years it had not been identified. It is a 'crouching giant' revealed only by the tip of its 'iceberg'.

If there are so many LSB galaxies, they may cover wide areas of the sky, and this means that they will block out the light from very remote objects such as quasars. This has been confirmed by quasar spectra; there are 'dips' in these spectra, showing that there is something in the way. There are too many of these 'dips' to be caused by visible galaxies, so that LSB systems are presumably responsible. On the other hand, there must be an upper limit to the number of such systems, as otherwise the whole sky would glow. Disney estimates that the total amount of possible 'galaxy light' cannot be more than 5 to 10 times the amount observed at the moment.

The compact galaxies are different. To search for them manually would be hopelessly time-consuming, and again a special machine is used. After scanning, say, 100,000 stars, it may be expected that a few new galaxies will be identified. Some of them look like stars, and are close and compact; others may be very distant and luminous, perhaps of the Seyfert variety.

But there is another revelation. Not very many high surface brightness galaxies (HSBGs) are found, because apparently the light we receive comes from the surface regions only; the small black clouds of 'smoke' and dust seen even in normal galaxies indicate that there is a great deal of obscuration. Galaxies are not nearly so transparent as we have always believed.

With present techniques we may have done all that is possible. The need now is to extend the field of research. At Cardiff, Disney's team has built a special camera which I have called the 'Hitch-Hiker',* because it can be used on a large telescope during routine observations without causing any interference with the main programme. It 'looks over the astronomer's shoulder' while he is carrying out his own research, and it can obtain at least twenty deep-sky frames every night, many of which will show up objects of very low surface brightness.

If there is indeed a great deal of previously undetected matter in the universe, both in the form of LSB systems and in visible galaxies themselves, will this be enough to 'close the universe' and, ultimately, stop the expansion? It is possible. We must await further research, and the Hitch-Hiker is already playing a major rôle. At least we have shown that in studies of galaxies, observational selection is far more important than had been previously realized.

*I made this chance remark during the actual programme, and it was taken up; by now it has become 'official', so that at least I added a new term to astronomical language!

7 OBSERVATORIES OF THE ATACAMA DESERT

Which is the world's most impressive observatory? It is hard to say. Hawaii is spectacular; so is La Palma. But to my mind the three observatories high in the Chilean Andes, over 8000 feet above sea-level, are the equal of any. I had never been to them before January 1989, and I did not really know what to expect.

All three—La Silla, Las Campanas and Cerro Tololo—are within range of each other, and by now they contain some of the very best telescopes ever made. We flew to Santiago; thence to the little town of La Serena, by Chilean Airlines (which were surprisingly good) and then drove the rest of the way, which took several hours.

To say that the Atacama Desert is barren is to put it mildly. The temperature would have been about 105 degrees in the shade—if there had been any shade! We passed through a few small Indian villages—one couldn't call them 'towns'—but otherwise there was nothing, and I gather that apart from the scattered Indian settlements there are no living things apart from desert foxes. (One of these, incidentally, comes to La Silla every day for breakfast. I did my best to make friends with it, but, not unnaturally, it was coy.) The Atacama, by the way, is not sandy like the Sahara; it is stony.

The astronomers came to Chile for the excellent reason that seeing conditions are superb, and probably as good as any in the world. Of the three observatories, La Silla is run by the ESO (European Southern Observatory) and the other two by the Americans. We went first to La Silla, to be welcomed by Daniel Hofstadt, who is in charge of all the instrumentation.

It is a truly breathtaking site. There are fifteen domes in all, plus a 15-metre or 49-foot radio telescope dish used mainly in the sub-millimetre range. They are widely spread out, and the Observatory owns a large area—almost 250,000 square miles, which, believe it or not, was purchased in the 1960s for the princely sum of 10,000 dollars (I wonder what it is worth now?). Rainfall is very low. In past ages the area was apparently forested, with plenty of water; not far from the Observatory there are some rock drawings which have never been deciphered, but which appear to show 'waves' together with the Sun. Daniel Hofstadt commented that the unknown artists must have been among the first astronomers!

The seeing is magnificent, as I have said, but precautions have still to be taken. For example, the prevailing winds are from the north, so that unwanted heat is blown away to the south, and the various domes are aligned accordingly.

The modern practice is to give dimensions in metric rather than the Imperial measurements which everyone can understand. So I will defer to convention. The main telescopes at La Silla are the 3.6-metre (141-inch) reflector, the 2.2-metre (87-inch) Max Planck telescope, the 1.5-metre (60-inch) Danish telescope, and the new 3.5-metre (138-inch) NTT or New Technology Telescope, of which more anon. There is also the 1-metre (39-inch) Schmidt, which has been used for the Southern Sky Survey under the leadership of Richard West—who, despite his name and his impeccable English, is himself a Dane. (English is about

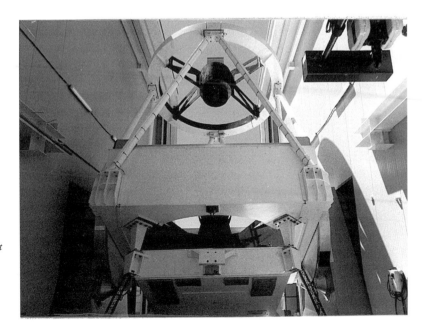

the seventh of his nine languages. It tends to give humble mortals such as myself an inferiority complex.)

The primary object of the Southern Sky Survey was to detect new features, and so far 15,000 new galaxies have been found, plus many clusters, comets and hundreds of asteroids. The programme has been carried out in conjunction with the UK Schmidt telescope at Siding Spring in Australia; generally the blue-sensitive plates are taken in Australia and the red-sensitive at La Silla. (I still regret that Britain is not a partner in the ESO; we pulled out at an early stage of planning, which I am sure was a mistake.) Of course the latest electronic devices are used, mainly CCDs (Charge-coupled Devices), but photography is not neglected, and with new emulsions is even staging something of a comeback. A new means of studying the plates has been developed at La Silla. Before being developed, the plates are placed in a tray above which is a horizontal grid. The grid oscillates, agitating the developer and making the overall performance much better than it would otherwise be. The problem is that there is only one millimetre clearance between plate and grid, so that if there is any error it is rather unfortunate for the plate!

Objects of special interest are then studied with the 3.6-metre telescope. When completed, in 1976, it was larger than anything in Europe, and it is still in the world's 'top ten'. There are various optical systems: prime focus, Cassegrain, and a third focus where a fixed spectrograph is placed. A smaller telescope, the Coudé Auxiliary Telescope or CAT, feeds light into the spectrograph. Very faint objects, such as quasars, are studied in visual and infra-red wavelengths. The telescope is mounted on the conventional equatorial pattern; Richard West commented that it had been nicknamed 'the last of the dinosaurs'—but it is a very effective dinosaur indeed. Naturally, the field of view is smaller than that of the Schmidt, which is 5 degrees square. The two telescopes work in collaboration, and the Sky Survey shows objects a hundred times fainter than anything seen before in the southern hemisphere.

So far we had seen a great deal—and were duly impressed. Then Richard West had another suggestion. Would we like to have a look at Halley's Comet, now on its way back to the depths of the Solar System and already beyond the orbit of Saturn? We needed no urging, and at nightfall we made our way to the Danish telescope, where Richard was already at work.

Though so much smaller than the NTT or the 3.6-metre, the Danish telescope is still a giant; after all, it is the equal of the first large instrument at Mount Wilson. But we were not to 'look through it'. Today everything is computerized, and in any case we would not have been able to see the comet, now down to magnitude 24½, and therefore sixty million times fainter than the dimmest object visible with the naked eye. Richard commented that seeing it was the equivalent of picking out the Little Mermaid statue in Copenhagen if it were illuminated only by sunlight.

The first step in an operation of this kind is to locate the field. To make matters more difficult, the comet is moving all the time with respect to the stars—and the telescope has to be guided so as to follow a comet which, initially, cannot be glimpsed at all. A CCD is used, of course, and the integration takes half an hour. We experimented first with a brighter comet, Wilson's, and then began to look for Halley.

As the picture was unveiled on the television screen, we first saw the streaks which represented the stars—and then, to our delight, the fuzzy blob which was the comet. I remembered seeing it from Australia, in 1986; it then had a tail, and was quite conspicuous with the naked eye. Now it had no tail, but it was clear that activity was still going on even at the comet's immense distance from the Sun. It was patently unstellar. I was glad to see it—after all, it might well be my last chance; I doubt whether I will be around when it is picked up again prior to the return of 2061!

At the time of my first visit the New Technology Telescope was still in the 'testing' stage. When I next went there, in 1990, it was in full operation, and had shown that it is probably the most effective telescope in the world today. Certainly it is of revolutionary design; at first sight it does not look like a telescope at all.

Its building, perched upon a hillock atop La Silla, looks most extraordinary. Instead of being a graceful dome, it is square and silver; it moves round as the telescope moves, so that the slit is always in front of the telescope itself. The NTT can move only in declination (that

Centre of the Tarantula Nebula: (30 Doradûs) in the Large Cloud of Magellan, photographed with the NTT (New Technology Telescope) at La Silla. Reproduced by permission of the European Southern Observatory. The picture shows the central cluster; in the middle is an apparently unresolved object which is in fact a compact cluster of very hot stars. Exposure time, 1 minute. 19 December 1989.

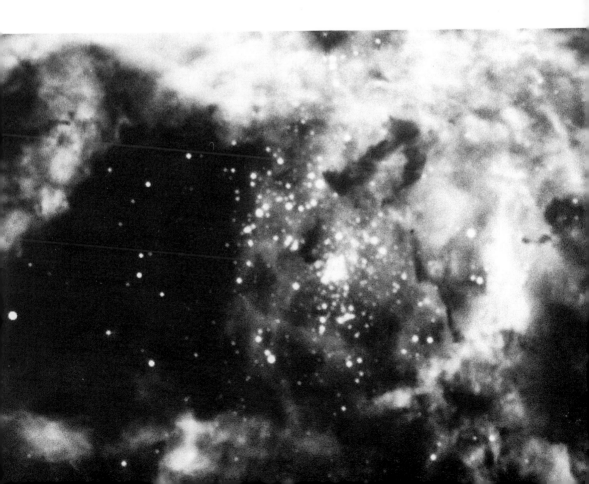

is to say, up or down), because the east-west movement is dependent upon the observatory building. The mounting is, in fact, a type of altazimuth. The telescope has to be driven with two computers instead of one, as with the equatorial, but nowadays this is no problem at all, and the old-type equatorial mounting is fast becoming obsolete insofar as new giant telescopes are concerned.

The heart of any telescope is its main mirror, and that of the NTT is as perfect as any ever made. It is very thin, with a maximum thickness of only 24 centimetres, and it weighs 6 tons; it is made of the low-expansion glass ceramic Zerodur. A classical mirror of the same size would weigh almost twice as much. The whole telescope 'floats' on a layer of oil only 0.03 of a millimetre thick; the oil temperature is controlled to within a tenth of a degree, so as to avoid producing heat which would cause disturbance in the surrounding air.

Swinging a large mirror around means distorting it, and this is particularly damaging with a very thin mirror. Therefore, two systems are used to compensate for any distortions. The first is termed active optics, i.e. altering the shape of the mirror so that it always retains its perfect curve. The way to do this is to have computer-controlled 'pads' behind the mirror, so that they can make the compensations automatically. The NTT has 78 of them. It has been found that 80 per cent of the light from a star can be concentrated into 0.125 of a second of arc, which is three times better than can be managed by any other ground-based telescope of equivalent size.

Active optics were brought into use in 1989, and proved to be very successful, but 'adaptive' optics are more of a problem, because they aim to remove the distortions due to rapid changes in the Earth's air. This means that a small, computer-controlled mirror is inserted in the telescope, in front of the light-sensitive detector. By monitoring the image of a relatively bright star in the field of view, the mirror's shape is continuously modified so that the distortions due to air turbulence can be removed. Adaptive optics were tested with a prototype instrument in France, in the infra-red region of the electromagnetic spectrum, and worked well. They will be fitted to the NTT in the very near future—perhaps even before this book appears in print.

Another development is that the NTT can be operated by remote control. There is a control room at Garching (near Munich), the head-quarters of the European Southern Observatory, and this is compatible with the control room in the NTT building, so that the observer has no need to make the long and expensive journey to Chile; he can operate the telescope from Garching—and this has the additional advantage that an observing programme can be tailored to fit the prevailing conditions at La Silla.

'First light' with the NTT was obtained on 23 March 1989, and it soon became clear that the telescope was every bit as good as had been hoped. One of the first objects to be studied was the globular cluster Omega Centauri. Globular clusters are huge, symmetrical star-systems, each containing up to a million suns, lying round the edges of the main

Galaxy; by cosmical standards they are relatively old, and because they are so remote not many of them are bright. Omega Centauri is much the most conspicuous of them, and is easily visible without optical aid; it was recorded by Ptolemy as long ago as the second century AD. (The only other globulars definitely visible with the naked eye are 47 Tucanæ, in the area of the Small Cloud of Magellan, and Messier 13 in Hercules.)

European observers never cease to bemoan the fact that Omega Centauri is so far south. Its distance is around 17,000 light-years, and the rich central core is about 100 light-years across; in this region the average distance between stars can be no more than about a tenth of a light-year, so that if our Sun lay in the middle of a globular cluster there would be no proper darkness at all. There would be dozens of stars shining brightly enough to cast shadows, and many of these would be old Red Giants.

Even a small telescope will resolve the outer parts of Omega Centauri, but the core is a different proposition, and until the advent of the NTT the individual stars there could not be seen separately. Now they can; the pictures obtained speak for themselves.

Another early NTT picture showed what has become known as the Einstein Cross, a classic example of a 'gravitational lens'. The story began with Albert Einstein himself (who else?) who predicted theoretically that a beam of light passing close to a massive body will be bent. Proof came in 1919, when images of stars near the totally eclipsed Sun were found to be displaced by just the amount which Einstein had predicted. But a gravitational lens is different. In this NTT picture, there are four images of a background quasar—because the single quasar has had its light split up, and there are five images altogether. On this unprocessed view, the central galaxy looks diffuse.

Of the other instruments at La Silla, special mention should be made of SEST, the Swedish sub-millimetre telescope, and also a fine 16-inch refractor, which was not being used for any official programme while I was there; I was able to turn it toward Saturn, and see the strange radial 'spokes' in the bright ring which are decidedly elusive except in the space-craft pictures. Nightfall at La Silla is awe-inspiring. You can see the Earth's shadow sweeping across the bleak landscape, so that in a matter of minutes the daylight has faded into velvet blackness.

New equipment is being planned. In particular there is the VLT or Very Large Telescope, which will not be a single instrument, but will be made up of four 8-metre (315-inch) telescopes working together—yielding light-gathering power equivalent to a single 630-inch mirror, more than three times the diameter of the Hale reflector at Palomar. Such an array provides great flexibility. For part of the time the four telescopes will be combined; at others they may be used separately. But the VLT will not be erected at La Silla itself. It is going to Paranal, well to the north though still part of the Atacama Desert. Paranal is very difficult to reach at the moment, but it does have the advantage of being just about the driest place in the world; no rain has ever been

La Silla, the Chilean station of the European Southern Observatory.

known to fall there—and moisture is the main enemy of the infra-red astronomer.

Of course, La Silla itself is a long way from any town, and there is relatively little light pollution. Isolation has its drawbacks; food is no problem, because the road from La Serena is good, but for maintenance of the equipment the astronomers have to rely upon the resources of the Observatory itself. Altogether there are about a hundred and fifty people running the technological and administrative departments. There are also ten scientific staff members, whose main function is to help visiting astronomers to become used to the various instruments. After all, no two telescopes are exactly alike. Not many visitors are disappointed, because bad weather at La Silla is rare; there are between 80 and 85 per cent of nights which are usable for observing, and around 80 per cent which provide what is called 'photometric' seeing, with more than six hours of continuously good conditions. This ratio is good by any standards, and I doubt whether it can be equalled anywhere else in the world except, possibly, the summit of Mauna Kea in Hawaii.

At the end of that first visit I left La Silla with great regret. It has a limitless future ahead of it—the site was indeed well worth the ten thousand dollars paid for it less than thirty years ago! Meanwhile, our next port of call was the second of Atacama's major observatories, the American-run Cerro Tololo.

Driving up the road, you come across Cerro Tololo quite suddenly; it is more 'condensed' than La Silla, and all the main domes are set up upon what looks like a smallish plateau, with the living quarters below and nearby.

The largest telescope is the 4-metre or 158-inch reflector, which at the time when I first visited Cerro Tololo was actually the largest telescope in the southern hemisphere. The mirror is truly enormous; it is two feet thick, and weighs 15 tons. It has a very short focal ratio (*f*/2.8) which is one reason why the telescope is so versatile. Needless to say, it is no easy matter to make a mirror of this sort; the construction time was between two and three years.

The mounting is of the conventional equatorial type. The telescope—all 375 tons of it—floats on a thin film of oil at a pressure of well over 100 pounds to the square inch, and yet the telescope is so delicately balanced that you could push it around with one finger.

For direct photography, the observer sits at the prime focus in a cage near the top of the tube. There is also what is termed a Ritchey-Chrètien focus, which is along the same lines as a Cassegrain, with the light-rays reaching the observer via a hole in the main mirror. There used to be a Coudé room, but this is now disused; the starlight is directed into the spectrograph by means of the newly developed principle of fibre optics.

When I arrived at Cerro Tololo, the 4-metre was being used by Dr Nicholas Suntzeff for a particularly fascinating programme. 'We're taking plates of a region in the Large Magellanic Cloud,' he told me, 'to study the proper motion of the Cloud itself. Actually, the Cloud was one of the first objects to be studied when the telescope was originally

Supernova remnant in the Large Cloud of Magellan (N49). Photograph taken in red light on 6 January 1990 by H. Pedersen at La Silla, Chile. Reproduced by kind permission of the European Southern Observatory.

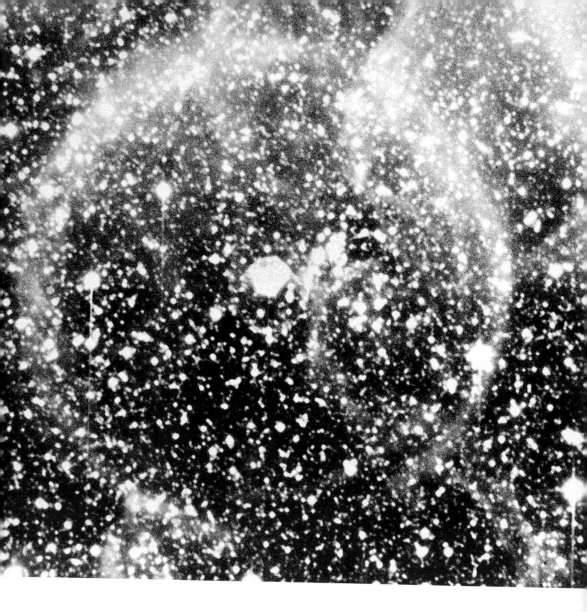

*'Light echo' from the
1987 supernova in the
Large Cloud of
Magellan. The ring is
due to light from the
supernova being reflected
from dust-clouds in
front of it. Photograph
by M. Tarenghi, La
Silla. Reproduced by
kind permission of the
European Southern
Observatory.*

brought into use in 1970. Since then, the stars in the Cloud have shifted slightly, and this ought to lead us on to a value for the transverse velocity of the whole system. Because the Cloud is 169,000 light-years away, the proper motions are incredibly small, and the probable value is no more than 6 milli-arc seconds per century. The reason why we can detect such a small shift is that we are measuring literally hundreds of thousands of stars relative to an absolute reference frame of galaxies and quasars behind.'

Of course, the background objects are so remote that their own proper motions can be ignored; the Cloud itself is a member of the Local Group of galaxies, and is cosmically almost on our doorstep. All the same, the tiny shifts are excessively difficult to measure. When I asked Nicholas Suntzeff whether anything definite had been found, he replied: 'We think so. It's a curious result. The Cloud appears to be moving in a direction exactly opposite to that which we'd expected. The two Clouds make up what may be called a binary galaxy system, and certainly both are enveloped in a cloud of rarefied hydrogen gas. This gas-cloud tails off behind the Large Cloud into what is termed the Magellanic Stream, and the "head" of this is apparently in front of the Small Cloud or SMC. You would expect both the SMC and the LMC to be moving away from the "tail", but in fact the LMC seems to be moving toward the "tail", which is startling. At the moment it can't be explained. If it is confirmed, it is going to make our understanding of the two Magellanic Clouds rather complicated.'

The Local Group also contains many dwarf systems, such as the small galaxy in Carina. Using radio methods at Cerro Tololo, it has been established that most of the mass of the whole Local Group comes from material which cannot be seen, and is presumably non-luminous; in fact, only about 10 per cent of the total mass comes from visible objects.

The 4-metre is the largest telescope at Cerro Tololo, and from the catwalk I could see all the others: the 1.5-metre multiple purpose telescope, the 91-cm telescope set up specially for use with the new electronic devices, the 41-cm telescope used mainly with photometers and photomultipliers, the 1-metre Yale telescope, a smaller Schmidt, and a 1-metre radio 'dish' which, unusually, is itself enclosed in a dome (most radio dishes stand proudly in the open). Further down the hill is the smaller Lowell telescope, used mainly for photoelectric photometry. In the near future—perhaps as early as 1995—there will be a new instrument, made up of four 4-metre mirrors working together to yield a light-grasp equal to that of a single 8-metre mirror.

One other device which intrigued me was something which I described as a 'bleeping bucket'. It is in fact an echo sounder, sending sounds into the air to measure the thickness and state of the atmosphere!

We drove down the winding road from Cerro Tololo on our way to the third of the great Chilean observatories, Las Campanas, where we were greeted by the resident astronomer, Dr William Kunkel. Here the main telescope is the 100-inch or 2.5-metre Irénée Du Pont reflector—which is exactly the same size as the famous Hooker reflector on Mount Wilson, which was for so many years not only the largest telescope in the world, but in a class of its own.

The Du Pont telescope has a very large secondary mirror, as it is intended to work over a very wide field—over one degree in diameter—making it the world's largest photographic telescope in the sense of taking plates 20 inches or 50 centimetres square.

'Perhaps the most interesting work going on now,' William Kunkel told me, 'is a major survey attempting to discover the nature and extent of the distribution of galaxies on a large scale. We are particularly interested in what has become known as the Great Attractor, which could well be a collection of galaxies much more massive than the famous Virgo cluster of galaxies some 60,000,000 light-years away. The Attractor is affecting the expansion of our own particular part of the universe. The Hubble constant—which tells us the rate at which the universe

William Kunkel, Director of the Las Campanas Observatory in the Atacama Desert of Chile.

is expanding—has been found to be less constant than expected; it is rather 'rubbery', so that there are different values of it in different regions of space.'

The whole problem of the Great Attractor has been widely discussed, and we are still not sure whether there is any Attractor or not! But Las Campanas is well suited to this sort of research, not only because of the superb observing conditions but also because astronomers can come and have full use of a major telescope for a set period, usually a couple of weeks, so that they can obtain all their data and then analyse it at their leisure.

There is also a 1-metre or 39-inch telescope named in honour of Henrietta Swope, who financed it, and there is a 10-inch telescope, in a run-off roof observatory, which was used to make a very important—if fortuitous—discovery in February 1987. William Kunkel told me the story:

'It was one of the most unusual collaborations in international astronomy. The episode began with Drs Alan Sandage and Gustav Tammann, who had a joint project of searching for supernovæ in external galaxies. This involved using thousands of photographic plates, and the aim was to cover a very large portion of the southern sky to hunt for supernovæ in clusters of galaxies; they expected something of the order of 25 supernovæ in two years. Among the plates purchased was one defective consignment. Each of the plates had a curious and devastating fault; for every square inch of plate surface there were around a thousand very tiny black dots, looking just as supernovæ would be expected to look. Obviously they had to be rejected, and they were simply stored away.

'At that time Ian Shelton, a Canadian astronomer, was here in charge of the 24-inch telescope funded by the University of Toronto. He is very devoted to astronomy, and as well as his official work he was anxious to pay attention to Halley's Comet, which passed perihelion in early 1986. The 10-inch telescope was available, and since a comet is not a point source the defective plates would be quite usable—so he agreed to take them. In his enthusiasm to photograph the comet, he wanted to study it as close to the Sun and under the most difficult conditions possible, so among the plates which he took were some obtained just before the comet passed behind the Sun and just after it emerged on the other side. In fact the comet was circumpolar, and very low in the sky. What better test objects could he have than the Magellanic Clouds? So he photographed the Clouds when no sane astronomer would do so—when they were very low over the horizon. Two nights before the supernova flared up, one night before, and on the actual night, he took a sequence of three plates of the Large Cloud which led to the discovery.'

At Las Campanas, the supernova was also seen visually by Oscar Duhalde. Within hours it was being intensively studied from every

Oscar Duhalde, of the staff at the Las Campanas Observatory.

observatory from which it could be seen, and it ranks as one of the most important astronomical events of the twentieth century; after all, it is the first naked-eye supernova to have appeared since Kepler's Star of 1604.

Las Campanas, too, has plans for an 8-metre or 315-inch telescope, but this will have a single mirror, produced by spinning the material on a turntable to give it its parabolic shape. It will be what Kunkel called 'the deepest mirror to be built, with a focal ratio of 1 : 2, so that the telescope will look rather like a box. It will have a curious mounting inasmuch as there will be no bearings at all. The altitude motion will be achieved by floating on a couple of rings, and the whole instrument will sit on a round, rotatable table. The dome will be relatively small. Two sites have been tested, both close at hand; we hope that the telescope will be ready some time between 1995 and 1998.'

Las Campanas is much the smallest of the three great Chilean observatories, but the quality and the exent of the research carried out there is the equal of any. Quite apart from this, Las Campanas is a friendly place—even though it is so isolated in the wildest part of the Atacama Desert.

Television from the summit of Mauna Kea, in Hawaii! I am facing the camera. . .

8 THE INTERNATIONAL ULTRA-VIOLET EXPLORER

In January 1978 a most remarkable satellite was launched. Its full name is the International Ultra-violet Explorer, though it is more commonly known simply as the IUE. It was scheduled to continue operating for a maximum of three years, and even this was widely regarded as optimistic; in fact it was still operating excellently in 1992, and has provided more material for research papers than any other satellite in history.

The ultra-violet region of the electromagnetic spectrum is extremely rich, but not a great deal of work can be carried out from the Earth's surface, because the ultra-violet radiations from space are blocked by layers in the upper atmosphere. For this reason the IUE was put into an orbit which takes it between 16,000 and 29,000 miles above the Earth, with a revolution period of 24 hours. This orbit was also suitable for operational purposes, because the satellite is always in view from one of its ground stations, which are situated at Goldstone (United States) and Spain (not far from Madrid). However, it does mean that the IUE is always in sunlight, and if the Sun shone down on to the sensitive ultra-violet telescope there would be serious damage. To make sure that no scattered sunlight edges into the wrong place, a system of baffles has been installed. This was one feature of the IUE which could not be tested before launch, but in the event the baffles have worked perfectly. The 'forbidden area' extends to 40 degrees away from the Sun; this value was chosen because of the wish to include Venus among IUE's targets.

Let us now go through an observing sequence. First, the proposal has to be vetted and accepted. The observer then travels to his ground station; the first target is selected; the co-ordinates are fed into the computer, and the IUE is slewed until the field has been identified. It is then kept firmly in place. The computer begins to feed the picture into the spectrometer, and after the end of the exposure (which could take two or three hours) the image is relayed in the form of an echelle spectrogram. The echelle principle involves a single spectrum which is optically cut up and 'stacked', so that it becomes manageable in the form of a two-dimensional picture; for study, the strips can be 'unstacked' and the spectrum examined in the usual way.

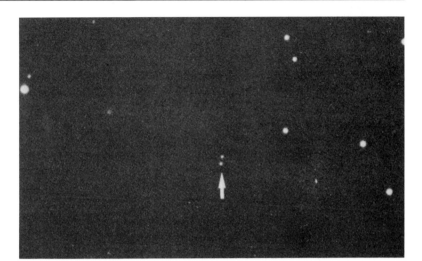

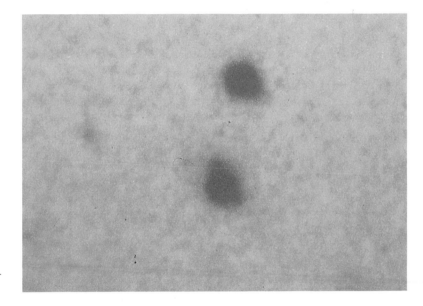

Double quasar, 0957+561: The two images are of the same quasar; the image has been split because the rays of light pass near an intervening galaxy— not visible here.

The longevity of the IUE has been remarkable. Of course, there have been hitches. Early on one of the computers acted in a schizophrenic fashion, and persistently tried to slew the IUE in the direction of the Sun, so that it had to be cut out; the problem was subsequently found to be one of heat—at a certain critical temperature, the computer started to misbehave.

All sorts of objects have been studied from the IUE. These include most members of the Solar System, apart from the Sun (too bright), Mercury (too near the Sun) and Pluto (too faint). Of special interest is Jupiter, with its volcanic satellite Io. Material from Io is ejected in the form of sulphur dioxide. This is not affected by Jupiter's magnetic field until it is ionized by collision, but then it is swept along, and the magnetic field moves faster than Io itself, so that the end result is a doughnut-shaped torus centred on Io's orbit.

Further out, great attention has been paid to stellar winds. It has been found that all stars show them (that of the Sun, of course, has been known for a long time, though it must be classed as a mere breeze by stellar standards). The strongest winds come from very luminous, unstable stars, such as those of the P Cygni type and the Wolf–Rayet type. Radiation pressure causes material to be blown away, and the loss may be as much as one solar mass every 100,000 years or so. Yet radiation pressure cannot be the only cause; there must also be another factor, whose nature is as yet uncertain. Then there are binary systems, notably those in which one component is a compact object such as a neutron star with an escape velocity of around 30 per cent of the speed of light. Studies have also been made of the hot haloes of galaxies. Our own Galaxy has a wide halo with a temperature of some 200,000 degrees,★ so that it is best studied at short wavelengths. Moreover, the haloes of galaxies imprint absorption lines on the spectra of background quasars, providing information about the intervening intergalactic matter.

Quasars themselves have come under careful scrutiny. Of special interest is the 'lens effect', where a single quasar appears double or even multiple because of the gravitational influence of a galaxy lying almost in front of it. In one famous case there are two quasar images which are not identical inasmuch as one of them is slightly distorted; when the effects are combined, the culprit is found—a giant elliptical galaxy lying almost in the mid position. The main importance of this is in the 'time delay' effect. Because the alignment is not exact, the light producing one image will have a slightly shorter distance to travel than with the other image, and so will be seen slightly 'earlier'; this is Image A in

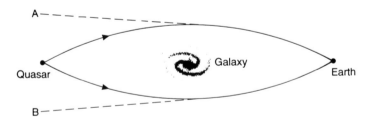

★This does not mean that the Galaxy is encircled by a sort of cosmic oven. The scientific definition of 'temperature' depends upon the speeds at which the atoms and molecules move around; the faster the movements, the higher the temperature—but with very rarefied material, there are so few particles that the amount of 'heat' is slight. The classic example is that of a firework sparkler compared with a glowing poker. Each spark has a far higher temperature than the poker, but has so little mass that it is safe to hold the firework in the hand—whereas I for one would hate to grasp the end of a red-hot poker!

the diagram. When the quasar temporarily faded (as it conveniently did, soon after it was identified), Image A showed the decrease first, and Image B did so only 21 months later. This made it possible to work out the distance of the background quasar, which proved to be around 10,000 million light-years. It also gave new information about the rate of expansion of the universe, and a re-determination of the Hubble constant at about 70 kilometres per second per megaparsec. With this particular quasar, we are looking back to a time when the universe was only about 27 per cent of its present age.

The IUE has been a phenomenon. Eventually its solar panels will be put out of action by particle bombardment, but it is safe to say that this gallant little satellite has been one of the major successes to date.

9 NEW PROBLEMS OF PLUTO

In the summer of 1989 I visited the Lowell Observatory at Flagstaff, in Arizona. I was no stranger to it; many years before I had spent a good deal of time there, using the large refractor to map the Moon before the Apollo astronauts went on their epic journey. This time I was merely 'passing through' and I went to see a very famous telescope: the 13-inch refractor which is officially known as the Lawrence Lowell Telescope, but which is best known as the instrument which took the photographs identifying the ninth planet, Pluto.

The 13-in Lawrence Lowell refractor at Flagstaff, used by Clyde Tombaugh in his successful search for Pluto.

This was also an appropriate time so far as Pluto is concerned. Its orbit is much less circular than those of the other planets, and when at its nearest to the Sun it is considerably closer-in than Neptune. It reached perihelion at just about the time I went to Flagstaff. Between 1979 and 1999 Neptune, not Pluto, ranks as 'the outermost planet'.

Yet is Pluto a proper planet at all? This is debatable. In fact, ever since its discovery in 1930 Pluto has set astronomers puzzle after puzzle, and even now our knowledge is depressingly meagre. Moreover, it remains the only 'senior' member of the Sun's family not to have been contacted by a space-craft.

Pluto's story really goes back to 1781, when William Herschel discovered the planet we now call Uranus. When the orbit of Uranus was calculated, it soon became clear that some unknown force was pulling it out of position. Two mathematicians, John Couch Adams in England and Urbain Le Verrier in France, independently came to the conclusion that this force was the pull of an undiscovered planet moving round the Sun at a greater distance. In 1846 the new planet, Neptune, turned up almost exactly where Adams and Le Verrier had predicted, and once more the Solar System seemed to be complete.

But was it? One man who did not think so was Percival Lowell, founder of the Flagstaff Observatory. Today we tend to remember Lowell mainly because of his admittedly wild theories about intelligent canal-builders on Mars, but this is unfair in view of his genuine achievements; he was a great benefactor of astronomy, a brilliant popularizer, and an expert mathematician as well. Using much the same methods as Adams and Le Verrier had done, he calculated the position of a ninth planet, basing his work mainly on the perturbations of Uranus—whose orbit was better known than that of Neptune, because Neptune, with its revolution period of over 164 years, had not completed a single journey round the Sun since its discovery. (In fact, it has not done so even yet.) He failed to find it, but in 1930, sixteen years after his death, the long-expected planet was found by Clyde Tombaugh, using the excellent little 13-inch refractor which had been obtained specially for the hunt. Lowell himself had failed simply because the new world was so much smaller and fainter than he had expected. It was most emphatically not a gas-giant similar to Uranus or Neptune, and even in powerful telescopes it appeared merely as a dot of light.

Naming was the first problem. T.J.J. See, an American who had once worked at Flagstaff, suggested 'Minerva'. But See was very unpopular with his colleagues, and the name finally chosen was Pluto, suggested by an eleven-year-old Oxford girl, Venetia Burney (now Mrs Phair; I talked to her recently, and she told me the full story). The name was apt enough. Pluto was the God of Darkness; his planet, so remote from the Sun, must be a gloomy place.

Problems began at once. For one thing, there was the exceptional orbit; for 20 years of its 248-year revolution period Pluto's distance from the Sun is less than that of Neptune, though the orbit is tilted by as much as 17 degrees to the ecliptic and there is no danger of a collision on the line. More significantly, there was Pluto's size. Instead of being reasonably massive, it appeared to be no larger than the Earth, and as the measurements became better and better the derived diameter became less and less. Someone facetiously suggested that at this rate it would not be long before Pluto disappeared altogether.

Clyde Tombaugh, with the blink-comparator he used in his search for Pluto. He discovered Pluto in 1930; I took this photograph in 1980, on the occasion of the fiftieth anniversary celebration.

Mrs Phair (née Venetia Burney), seen here with me in 1987. It was Mrs Phair who suggested 'Pluto' as the name for the new planet.

It became painfully clear that Pluto was in fact too small and too lightweight to have any measurable effect upon the movement of a giant planet such as Uranus or Neptune; one might as well try to divert a charging hippopotamus by throwing a baked bean at it. There was an extra development in 1977, when photographs showed that Pluto was not alone; it was accompanied by a second body of roughly half Pluto's diameter. It was named Charon, after the sinister ferryman who took departed souls across the River Styx on their way to the Underworld.

Once two bodies are found to be affecting each other, their masses can be worked out. Charon provided major shocks. It is close to Pluto (the two are only 19,000 miles apart, centre to centre), and the revolution period is the same as the rotation period of Pluto: 6.3 days. Pluto and Charon are 'locked', so that to a Plutonian observer Charon would appear

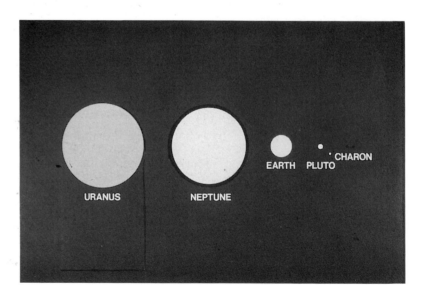

Apparent sizes of Uranus, Neptune, Earth, Pluto and Charon.

motionless in the sky—a case unique in the Solar System. And relative to Pluto, Charon is large. It is now known that Pluto's diameter is 1444 miles, considerably smaller than that of our Moon, while that of Charon is 753 miles. Certainly we are not dealing with a planet and a satellite— the pair gives a much better impression of a double asteroid—and there was no longer the slightest chance that Pluto was the planet for which Lowell and Tombaugh had been searching.

All sorts of theories were put forward. Pluto and Charon might really be asteroidal; there was a suggestion that they might be 'planetesimals', larger fragments left over, so to speak, when the main Solar System was being formed; it was even claimed that Pluto might be an ex-satellite of Neptune which broke free in some manner and declared cosmical UDI. But then, in the 1980s, planetary astronomers had two strokes of luck. Spectroscopic observations had indicated that Pluto had a thin

Pluto and Charon. (left) The best ground-based picture; (right) the image obtained from the Hubble Space Telescope; Pluto and Charon are shown separately. (below) Orbit of Charon.

Ground Based   HST/FOC

Pluto

Charon

atmosphere, and in 1988 this was confirmed when Pluto passed in front of a star, hiding or occulting it; both before and after the actual occultation the star's light had to come to us via the atmosphere of Pluto, and it was found that the atmosphere was unexpectedly extensive, reaching out to 370 miles from the surface. Also, it was decidedly hazy in its lower layers. Though Pluto is much less massive than the Moon it is also much colder, which is why it can retain at least a tenuous atmosphere.

The second stroke of luck occurred because during the late 1980s the orbits of Pluto and Charon were so placed, relative to the Earth, that there were mutual eclipses and occultations; Pluto's axis of rotation is tilted to the plane of its orbit by 122 degrees (another oddity). During 1987 and 1988 there were total eclipses of Charon by Pluto, and transits of Charon across Pluto, so that it was possible to study the two bodies separately—a state of affairs which will not recur for 124 years (half Pluto's revolution period). For the first part of the series Charon blotted out Pluto's north pole, then its equator and finally its south pole, providing astronomers with the chance to draw up a rough map. It seems that there is a dark band across the mid-southern latitudes of the planet, with a large, bright south polar cap.

It was found that Pluto showed traces of methane ice on its surface, while Charon showed water ice—but then, in 1992, Tobias Owen and his team from the University of Hawaii detected nitrogen ice on Pluto. From this they conclude that most of the atmosphere is composed of nitrogen, while the remainder is methane—hardly a surprise, because the same sort of atmosphere is found with Triton, the senior satellite of Neptune, which is slightly larger than Pluto.

The atmosphere may not always be present. When the planet moves out to aphelion the temperature drops, and the current atmosphere may condense out on to the surface as frost. There may be long periods when the atmosphere is even thinner than it is now, or even absent altogether.

What is Pluto really like? It may be not too unlike Triton, which was studied from Voyager 2, but we will have to wait for a space-probe to tell us more. At any rate, Pluto and Charon must be lonely and desolate beyond belief. The Sun would appear as a brilliant star, but few planets would ever be seen, and even Neptune at its best would cut a poor figure in the Plutonian sky.

10 THE MAKING OF AN ASTRONOMER

Over the years I have had countless letters, mainly from young enthusiasts, all asking the same question: 'If I want to become an astronomer, how do I go about it?' I usually try to give full and detailed answers, and there are several important points to be borne in mind straight away.

First, the professional astronomer must have a science degree—and there is no short cut. Qualifications are all-important in the modern world, and without a degree there is no hope of a worth-while post in astronomy or any other science. Amateur astronomy is different. No qualifications are needed apart from interest and enthusiasm; moreover, astronomy is one of the very few sciences in which amateurs can do useful work. For the moment, however, let us concentrate upon those who—unlike myself—have become professionals.

The University of Birmingham Observatory.

Modern astronomy is essentially mathematical. My advice: If you have no more than limited mathematical ability, then forget about astronomy as a career! It is also true that most professionals (admittedly, not all) are concerned with astrophysics, and this means that physics itself is crucial.

My usual advice to an inquirer is to recommend a basic degree in physics, probably with an astronomical bias. This gives a broader base than a degree in pure astronomy. Naturally there are exceptions, but as a general principle I believe that this advice is sound.

Most universities offer courses for would-be professional astronomers. One of the best is at Birmingham, where the Professor of Astronomy and Space Research is Peter Willmore—who, with Dr Ken Elliott and numbers of students, joined me for a special *Sky at Night* programme in September 1989.

School examinations come first. Physics and mathematics are essential, and it is useful, though not obligatory, to have another A-level as well. Armed with this, the student can begin his real work.

At Birmingham, one of the first steps is to go to the Astrolab, where there is a splendid collection of photographs taken at great observatories such as those at Palomar, Australia and Chile. The student is introduced to some statistical work, such as studying the numbers and distribution of stars in globular clusters; this may sound tedious at first sight, though it is a great deal more interesting than might be thought! Then, in the laboratories, students design and test their own equipment. It is remarkable how much original work can come out of this.

Obviously, use of practical telescopic equipment is needed. Birmingham University has its own observatory, some miles away from the city itself so as to obtain relatively dark skies. The main telescope is a 16-inch (0.4-metre) reflector. This is comparable in size with the reflector at my modest observatory in Selsey, but the Birmingham instrument is different inasmuch as it has been designed specifically for student use and training. It is equipped with an excellent spectrograph, and also with a CCD. This means that it can be used in just the same way as a giant telescope. Because of its relatively small aperture, it cannot reach faint objects, but this does not matter; it is the principles which count. A student who can handle the Birmingham telescope and its accessories will be equally competent at managing a large telescope at a major observatory.

Later there are special projects, in which groups of students band together in some particular piece of work. Again, original research can emerge. And immediately upon graduation, or even before, students are introduced to Starlink, the network of linked computers which now extends to observatories in most parts of the world. From Birmingham, it is possible to interrogate Starlink computers in places as far afield as Chile and Hawaii.

Quite apart from all this, Birmingham University is active in space research. Equipment built here has been flown in Spacelab missions, and although there have been serious delays in the overall programmes

Professor Peter Willmore, seen here with me in 1992.

(due largely to the *Challenger* disaster) there will be new missions in the future using Birmingham-built technology. In this, too, students play a major rôle.

Not all those who take the Birmingham course will become full-time astronomers, and indeed most of them will not. However, it is stressed that the main objective is to give the students a range of skills which will help them if they take up any other career, such as in industry or commerce. Knowledge of scientific methods is always invaluable.

Peter Willmore is optimistic about the future of British astronomy, and is quick to point out that the trainees of today become the senior researchers of tomorrow. The one weak link is the shortage of university posts, which is a fact of life. Let us hope that this problem can be resolved, because it is absolutely crucial.

I left Birmingham University feeling just as optimistic as Peter Willmore. So long as courses such as this remain popular, British astronomy is in good hands.

 **11 BULL IN THE SKY**

Winter skies are glorious. We have Orion, with all his retinue; Sirius has come into view; Capella, the yellow star of Auriga, is almost overhead. And we have Taurus, the celestial Bull, which contains objects of immense interest even though the constellation itself has no particularly distinctive outline. In mythology, it represents the bull into which Jupiter (or Zeus), the ruler of the gods, changed himself in order to carry off the beautiful maiden Europa. (What happened after that need not concern us here. It does not pay to inquire too closely into the morals of the ancient Olympians.)

To find Taurus, begin with Orion and follow up the line of the Hunter's Belt. Before long you will come to the bright orange-red Aldebaran or Alpha Tauri, the 'Eye of the Bull' and much the most brilliant star in the constellation. It is 100 times as luminous as the Sun, and is 68 light-years away, so that if you look at it now you are seeing it as it used to be at the time when Adolf Hitler was just starting to make his presence felt.

In colour it is much the same as Betelgeux in Orion, but Aldebaran is not nearly so powerful or so remote; it has a spectrum of type K, as against type M for Betelgeux. Aldebaran is almost steady in its light, whereas Betelgeux varies considerably. It is interesting to compare the two, though Betelgeux is nearly always the brighter by a few tenths of a magnitude.

Extending from Aldebaran there is a little V-formation of stars. These make up the Hyades, one of the best-known open clusters in the sky. The Hyades are easy to see with the naked eye, but the best view is probably to be had with binoculars. With a telescope not all of the cluster can be seen at any one time; it is too scattered. There are several hundred members in all, covering an overall diameter of about 40 light-years.

Oddly enough, Aldebaran is not a genuine member. It simply lies in much the same direction as seen from Earth, and is almost midway between the Hyades and ourselves, so that we are dealing with a line of sight effect; in a way this is rather a pity, because the bright orange light of Aldebaran tends to overpower the fainter stars of the cluster. From a different vantage point, Aldebaran and the Hyades could well appear on opposite sides of the sky.

There are two easy doubles in the Hyades. Delta Tauri makes up a wide pair with its neighbour 64 Tauri, and there is also Theta Tauri, which can be split with the naked eye. Binoculars show that the brighter

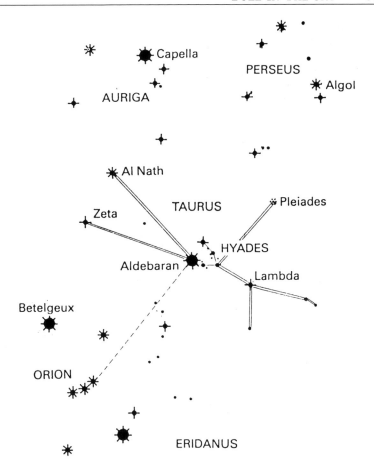

Orion and his retinue.

member of the pair (magnitude 3.4) is white, while the fainter component (3.8) is orange. Again we have a line of sight effect, because the white star is fifteen light-years closer to us than the orange companion. The two stars of Theta Tauri cannot be classed as a true binary system, though no doubt they, and all the other stars of the Hyades, were formed from the same nebulous cloud.

On the far side of Aldebaran, reckoning from Orion, we come to the Pleiades cluster, which is undoubtedly the most striking of its type in the entire sky. At first glance it looks like a hazy patch; closer inspection shows that it is starry. Many years ago, in a *Sky at Night* programme, I enlisted the aid of viewers in an attempt to find out how many individual stars can be seen with the naked eye by people of average eyesight on a clear night. The mean number was 7, justifying the familiar nickname of the Seven Sisters, but some keen-eyed viewers could make out more—as many as twelve in a few cases. The record is said to have been set by the last-century German astronomer Eduard Heis, who could count nineteen. Any binoculars will show dozens, and the entire cluster contains at least 500 members, with associated nebulosity.

This nebulosity is not easy to see visually, though it is not hard to photograph. It proves that the cluster is young by stellar standards, so that star formation is probably still going on. Moreover, the leaders of the Pleiades are hot and bluish-white; they have not had enough time to use up their main 'fuel' and evolve into Red Giants, as Aldebaran and Betelgeux have done.

The senior Pleiades have individual names: Alcyone, Electra, Atlas, Merope, Maia, Taygete, Celæno, Pleione and Asterope. Alcyone, of magnitude 3, is the brightest, and is at least 400 times as luminous as the Sun, so that it is much more powerful than Aldebaran. For some reason or other a famous nineteenth-century German astronomer, Johann von Mädler, believed Alcyone to be the central star of the Galaxy, with everything else revolving round it. I have never understood why he believed this. Mädler was a great observer, and was the first to draw up a really good map of the Moon, but as a theorist he was much less successful.

The Pleiades are 410 light-years away. If the Sun were a member of such a cluster we would have a glorious night sky, but there is no danger of the Pleiades stars crashing into each other; there is plenty of room to spare.

The Pleiades have been known since ancient times; they are referred to in the Odyssey, and also in the Bible. Inevitably, there are mythological legends attached to them. In one account the Pleiades were seven beautiful maidens who were strolling innocently in a wood when they were spied by the hunter Orion, who gave chase. It was fairly clear that his intentions were anything but honourable, and to save the maidens from a fate worse than death Jupiter changed them into stars and swung them up into the sky. Orion's views about the whole episode are not on record . . .

As with the Hyades, so with the Pleiades—the best view is obtained with binoculars, or with a very low-power eyepiece on a telescope. As soon as a higher magnification is used, the main beauty is lost, because the whole area of the cluster amounts to several square degrees, and cannot be fitted into the same telescopic field.

Next, come to the third-magnitude star Zeta Tauri; its proper name (Alheka) is hardly ever used nowadays. It is well over a thousand times as luminous as the Sun, and is a very active star, sending out 'shells' of material which we cannot see directly, but which can be detected with suitable equipment. (Pleione, in the Pleiades, is another shell star.) Close to Zeta Tauri is one of the most remarkable objects in the sky: Messier 1, the Crab Nebula.

Many books say that the Crab cannot be seen without a telescope. Admittedly it is far below naked-eye visibility, but with ×10 binoculars you can glimpse it, and with my 20×70 binoculars it is easy; it is also just in the same field with Zeta Tauri. A telescope shows it as a misty patch; photographs show that it is immensely complex. Its nickname was given to it by the third Earl of Rosse, who made a sketch of it in the 1840s with the great 72-inch reflector at Birr Castle in Ireland.

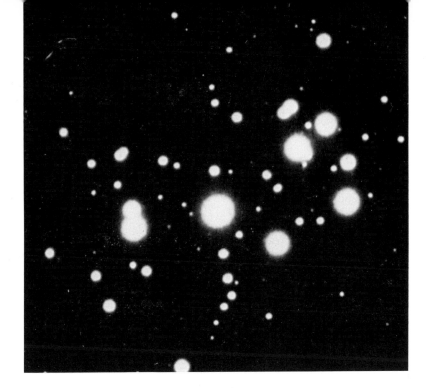

The Pleiades (Seven Sisters), the most famous of all open star-clusters.

The Crab is not a true nebula, but a supernova remnant. Supernovæ are colossal stellar explosions. One type involves the complete destruction of a small, dense star of the type known as a White Dwarf, but the Crab remnant is of the second type, produced by the collapse of a very massive star which has used up all its available 'fuel'. When energy production suddenly stopped there was an implosion, followed by a rebound, and the entire star blew up, shining temporarily with a power perhaps 100,000 million times that of the Sun. It took years for the fierce glow to fade.

The outburst was seen. On 4 July 1054 it was recorded by Chinese and Korean star-gazers, equalling Venus and visible in broad daylight; it remained on view for weeks. Unfortunately it is not well-documented, though some old Indian rock drawings from the south-west United States may show it.

When it sank below naked-eye visibility, it was lost until 1731, when the gas-patch marking its demise was detected by an amateur astronomer named John Bevis. Messier found it independently in 1758, and made it No. 1 in his famous catalogue.

The Crab is expanding, as we can see by studying past and modern photographs of it. It is 6000 light-years away, so that the actual outburst took place in prehistoric times. The remnant contains a pulsar, made up of neutrons, which is only a few miles in diameter, and is so dense that you could cram thousands of millions of tons of its material into a wineglass. The pulsar is spinning round 30 times a second, and as it does so it sends out the characteristic radio emissions. It is one of the very few pulsars to have been optically identified, and it is possibly the most intensively studied object in the whole of the sky. Since 1054 only two supernovæ have been seen in our Galaxy, those of 1572 and 1604,

The constellation of Taurus (the Bull).

though in 1987 we were privileged to see a supernova outburst in the Large Cloud of Magellan, a mere 169,000 light-years away.

Another star in the Bull worthy of mention is Lambda Tauri. To find it, use the Hyades as a sort of 'arrow-head' pointing toward it. Usually it is of magnitude 3.3, but every four days it gives a long, slow 'wink', dropping to magnitude 3.8 before brightening up again. Estimates made every few minutes near the time of minimum will show that it is changing; Gamma Tauri in the Hyades (magnitude 3.6) is a useful comparison. The times of minima can be predicted very accurately, and are given in periodicals such as the British *Astronomy Now*.

Lambda is not a true variable, but an eclipsing binary of the Algol type. It is made up of two components, so close together that they cannot be seen separately, moving round their common centre of gravity (actually they are about 8,500,000 miles apart). Every 3.95 days the fainter member of the pair passes in front of the brighter, and occults it by about 40 per cent, so that Lambda seems to fade. There is a secondary minimum when the fainter component is hidden, but the decrease is too slight to be noticed with the naked eye. Lambda is about 400 light-years away; both components are hot and white.

Finally, consider a star which the Bull has stolen! Auriga, the Charioteer, has a distinctive pattern; adjoining it is Al Nath, which is above the second magnitude, and which used to be known as Gamma Aurigæ. For some reason or other the International Astronomical Union, the controlling body of world astronomy, gave it a free transfer into Taurus, so that Gamma Aurigæ became Beta Tauri. To me this seems illogical, as Al Nath so obviously fits into the Auriga pattern, but the decision has been made.

In any case, with or without Al Nath, there is plenty to see in Taurus. The celestial Bull is one of the leading groups of our winter sky.

12 THE EPIC JOURNEY OF VOYAGER 2

Of all the interplanetary space-craft launched up to the present time, which has been the most successful? Opinions differ, but on balance I think that my vote would go to Voyager 2, which has provided us with almost all our detailed information about the two outer giant planets, Uranus and Neptune.

In the early 1970s it was recognized that the outer planets were moving into such a situation that they could be visited in turn by the same probe, using what is called the gravity-assist technique. Both the Voyagers were launched in 1977. Voyager 1 duly by-passed Jupiter and Saturn, while Voyager 2 added Uranus and Neptune as well. Of course, techniques were not then so advanced as they are now; the electronic devices known as CCDs were only just being tested, and were not nearly ready for a space mission. It is therefore all the more remarkable that the images obtained were so spectacular—and that Voyager 2 performed just as well at Neptune as it had done at Jupiter, so long before.

Voyager 2 completed its task with the Neptune encounter in August 1989. Therefore, this may be the right moment to look back at some of the highlights.

Jupiter, 1979—The thin, dark ring—so unlike the bright, icy rings of Saturn—was a surprise; but there were also the vivid colours of Jupiter's upper clouds, which told us a great deal about the Jovian 'weather systems'. Dr Garry Hunt, the planetary meteorologist, pointed out that the weather systems on Jupiter are driven very much like those on Earth, despite the tremendous amount of internal heat (Jupiter's core temperature is probably between 20,000 and 50,000 degrees Centigrade). We did not expect the winds or the turbulence to be so violent, but the main driving mechanisms were understandable enough.

The Great Red Spot is a whirling storm: a huge vortex, gaining much of its energy by sucking in smaller spots from time to time. The main mystery about it is its colour, so very different from the white ovals adjoining it. We do not yet know the full answer, and Jupiter's cloud chemistry is still very much of an enigma.

Of the satellites, Ganymede and Callisto turned out to be bright and icy, Europa icy and smooth. Io was striking, and is the most volcanic object in the Solar System. The basic energy of the volcanoes is certainly due to the tides raised by Jupiter. If Io were on its own, in a circular

orbit, there would still be tides, but there would be no changes; however, the other large satellites affect Io's motion and make the orbit somewhat eccentric, so that the tides do change, driving energy into the crust by friction.

There are several types of vulcanism on Io. There are ordinary volcanoes of the sort we find on Earth, with silicate lavas and calderas, but there are also 'plumes' extending to several hundreds of miles above the surface, which may be more akin to what we usually call geysers—driven by sulphur and sulphur dioxide encountering hot material below Io's surface.

To me, the most puzzling fact is that the difference between Io and Europa is so marked. Europa appears to be totally inert, and there are virtually no craters—certainly no volcanoes. We can only hope that we will learn more from the Galileo probe, scheduled to reach the Jovian system in 1995.

Saturn, 1981 (following the pass of Voyager 1 in 1980)—Here we found very high windspeeds, and were able to study the wind profiles from one pole to the other; they are very symmetrical, and seem to be driven by the same basic mechanism as with Jupiter, so that in this respect the two planets are alike. Magnetically they differ, because Saturn's magnetic axis is almost coincident with its axis of rotation.

The rings were much more complicated than had been expected, with thousands of 'strands'. Their stability is another problem. No doubt the satellites are involved in some way, but no 'shepherd' satellites were found in the ring system itself; possibly they are below the limit of Voyager's resolution (though much later, examination of the images revealed the existence of a very small satellite, now named Pan, moving in the Encke Division of the rings).

There were also the radial 'spokes' in the brightest ring, B, which should not be there at all; the particles closer to the planet move more quickly than those which are further out, so that any radial features should be broken up very quickly. Presumably they are associated with magnetic phenomena. Within the rings we now know that there are huge electrical disturbances, much stronger than any thunderstorms we experience on Earth, so that the spokes may be due to particles elevated away from the ring-plane by electrical forces.

The F-ring has several strands, which again are imperfectly understood. They show 'knots' here and there, and are unlike anything else we know.

Of the satellites, most are icy and cratered; Mimas has one vast formation, now named Herschel, which, if formed by impact, would have come very close to breaking Mimas up. Titan was found to have a dense atmosphere made up largely of nitrogen, but Voyager 2 obtained no close-range views; these had been sent back from Voyager 1—which is why the first of the space-craft was unable to go on to Uranus and Neptune; to image Titan, it had to be put into an orbit sending it well out of the ecliptic.

Saturn's C Ring, with the B Ring top left; Voyager 2, 23 August 1981, from 1,700,000 miles. The picture was obtained from images taken with the ultra-violet, clear and green filters.

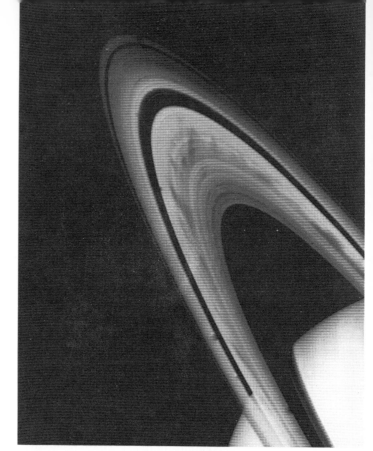

Spokes in Saturn's B ring: Voyager 2, 3 August 1981, from 14,000,000 miles. Each spoke is about 7500 miles long.

The dusty rings of Uranus, photographed from Voyager 2 at a range of 147,000 miles. The resolution is about 20 miles. The exposure time was 96 seconds, which is why the star images are drawn out into trails.

Uranus, 1986—The period following the Saturn encounter was very difficult for the Voyager team. Just before the Saturn pass was over, there was a serious problem with the scan platform which carried the cameras. For several years the engineers at JPL wrestled with it, until they finally found out the cause—lack of lubrication—and learned how to deal with it. Another hazard was the increasing distance, with consequent loss of signal strength. The capability of the DSN or Deep Space Network had to be increased; it was rather like trying to move Uranus in toward Saturn, or, later, Neptune in toward Uranus!

We had little idea of what to expect at Uranus, and the approach was unfamiliar inasmuch as it had to be 'pole-on'. Uranus has an axial inclination of 98 degrees, which is more than a right angle. Garry Hunt summed up the situation succinctly: 'All the time flying in, with Voyager getting closer and closer, I remember having numerous conversations with my colleagues about whether we would find any clouds in the Uranian atmosphere. Well, we found a few. Not many; but when we looked at the planet very carefully and tried to enhance the pictures, we saw that there were one or two large convective storms, and this gave us some ideas about the wind pattern. Having to look pole-on was very much of a complication. There seemed to be zonal motions of the kind we had seen before, but Uranus is peculiar; moreover it has no internal heat-source in the sense that Jupiter and Saturn have.'

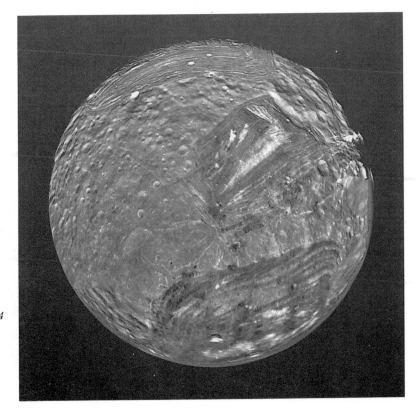

Miranda: Voyager 2, 24 January 1986—a full-disk, south polar view, showing the three main coronæ (Inverness, the rectangular feature Arden, and Elsinore).

The magnetic field is also strange, because the magnetic axis is offset to the rotational axis by almost 60 degrees, and does not even pass through Uranus' centre. Then there were the rings, which proved to be darker than coal-dust; it was found that there are hundreds of strands, so that the whole system, like that of Saturn, is very complex.

Ten new inner satellites were found, suggesting that there may have been many collisions of the type which replenishes the ring material. Of the previously known satellites, Oberon and Titania seem to be non-active geologically, though Titania shows evidence of faulting which may indicate past disturbances. Umbriel has some large craters on a darkish surface, and there is one brighter feature, now named Wunda, whose nature is uncertain; it may be a crater, but it lies right on the edge of the disk as seen from Voyager, so that our view of it is incomplete and highly foreshortened. Ariel has fault structures, with signs of icy crypto-vulcanism in the past. There are regions where craters seem to have been obliterated, and also features which look remarkably like glaciers.

Miranda is unique. According to Torrance Johnson, of the JPL imaging team, 'it requires some form of energy other than radioactive decay to explain the geology. Miranda is so small that it must have cooled down long ago.' Theorists have been studying the dynamics of the orbits of Miranda and Ariel; though they are not now 'in resonance', producing tidal energy in Miranda's globe, they may have been so in the past. Puck, largest of the new satellites, is cratered, irregular and very dark. 'To me,' said Torrance Johnson, 'these small satellites represent probably what a lot of the material that went into the satellites must have looked like four thousand million years ago.'

And so—on to Voyager 2's final target.

Neptune, 1989—What did we hope that Voyager would tell us as it sped past Neptune, skimming the darkened north pole from a distance of a mere 3000 miles? Well, we wanted to know the length of the rotation period; we were anxious to see whether Neptune could be a radio source; we hoped to decide upon the existence or non-existence of rings, and we wanted to study the satellites, of which two were known. Triton was of special interest because it was thought to have an atmosphere, and it is also unique, for a large satellite, in having retrograde motion, moving round Neptune in a sense opposite to that in which Neptune itself rotates. The second satellite, Nereid, was also interesting, but mainly because of its highly eccentric orbit, more like that of a comet than a satellite. New smaller inner satellites were fairly confidently expected. We hoped for good pictures, because Neptune gave every indication of being a more dynamic world than the bland Uranus, but we had to remember that Voyager 2 was an old probe, built in the 1970s, so that it did not carry modern-type electronic equipment. We hoped for the best.

Montage of the five main satellites of Uranus, from Voyager 2 images. From the top they are Ariel, Miranda, and Titania, Oberon and Umbriel.

There were strong contingents of planetary astronomers at JPL for the mission. I arrived early on, as I had been giving the opening lecture at a major NASA optical conference at San Diego, and there was no

*Ring arcs round
Neptune, as imaged
from Voyager 2.*

*Proteus, one of the
newly discovered
satellites of Neptune, as
imaged from Voyager 2.*

time to return home before setting off for Pasadena. Over the next few days other astronomers arrived, and then came the flux of journalists, science writers of all kinds, and the radio and television crews (in the event, I spent a good deal of time acting as a voluntary correspondent for CNN News). Before long, JPL was a hive of overt as well as covert activity. One visitor during the mission was Vice-President Dan Quayle, and I was able to meet him. Mr Quayle has had a great deal of unfavourable Press coverage, but I can only say that I was impressed with him; he was constructive, coherent, progressive, handled his news conference well, and had a pleasant personality too.

I had written the only current book about Neptune,* but during the mission my rôle was that of interpreter—and there was plenty to interpret. At an early stage it became clear that Neptune was much more active than Uranus. Pictures showing some surface detail had come through during the summer, and as Voyager closed in, during August, discoveries came quickly. Six new satellites were found, all much closer-in than Triton; Proteus, the largest of them, is actually bigger than Nereid, but is too close to Neptune to be seen from Earth. Nereid, alas, was in the wrong part of its eccentric orbit, and not much could be made out from the one picture obtained.

It is a tribute to the NASA engineers that as soon as the new satellites were discovered, adjustments were made so that two of them—Proteus and Larissa—could be imaged. Both turned out to be darkish and cratered, which was no surprise at all.

Next, the rings. It emerged that the so-called ring arcs, detected from Earth by occultation techniques, represented parts of complete rings, but even the main ring (now called the Adams Ring, after one of Neptune's co-discoverers) is 'clumpy'; its brightest segments had been mistaken for arcs. The segments also contain 'moonlets' a few miles across, though whether they are compacted bodies or mere swarms of particles is still uncertain. The faintest of the rings was only just above Voyager's threshold of visibility. *En passant*: after closest encounter, Voyager had to pass across the ring-plane, and this caused some alarm at JPL, because there was a brief period when the impact rate rocketed up to 300. However, Voyager 2 was unharmed, and the NASA planners heaved deep sighs of relief.

Next, the rotation was confirmed at just over 16 hours. This was by means of radio waves (I was actually in the main control room when these were discovered; it was quite a moment). The magnetic field turned out to be weaker than those of the other giants, and is inclined to the rotational axis by almost 50 degrees. Magnetically, therefore, Neptune is very like Uranus, and quite unlike Jupiter or Saturn; as with Uranus, the magnetic axis is considerably offset from the centre of the globe. Weak auroræ were detected, but these were of course in high magnetic latitudes, closer to the Neptunian equator than to the geographical poles.

*The Planet Neptune, Ellis Horwood Ltd, Chichester, 1989.

Just before closest approach, Neptune showed up in all its glory. The main feature was the GDS or Great Dark Spot, a huge oval at about latitude 22 degrees south. (Remember that because Neptune's southern hemisphere was having its long 'summer', this was the area best studied from Voyager 2; the north pole was in complete darkness.) The GDS has a rotation period of 18.3 hours, longer than that of the planet's core, and drifts westward relative to the adjacent clouds at almost ten feet per second. Moreover, it rotates in an anti-clockwise direction. Above it are bright, wispy, cirrus-type clouds made up of methane crystals, lying from 30 to 50 miles above the GDS itself with a clear layer in between. The clouds change so rapidly that it is difficult to keep track of individual features even from one rotation of Neptune to the next. It may well be that the true surface of the GDS contains hydrogen sulphide.

At 55 degrees south there is a second dark spot (D2) with a cloudy 'eye of the storm'; its rotation period is 16.1 hours, about the same as that of the core, and it laps the GDS every five Earth days. In between, at 42 degrees south, is the curious 'Scooter', with a quicker rotation period—hence the nickname.

From all this, we can work out the wind systems for the planet. Near the equator the winds blow westward, i.e. retrograde, at perhaps 250 metres per second. In the latitude of the GDS, the speeds increase to 325 metres per second (725 mph). Further south they slacken, ceasing at around latitude 50 degrees; at latitude 60 degrees south there is an eastward flow of up to 50 metres per second, declining again to zero at the pole. Since Neptune receives so little solar energy, these strong winds are surprising. They may be due to lack of turbulence, since turbulence 'wastes' energy—a situation very obvious on Jupiter, for example.

The upper atmosphere consists of 85 per cent hydrogen, helium accounting for most of the rest. Temperatures are much the same as for Uranus; Neptune's greater distance from the Sun is compensated for by its internal heat-source, lacking with its calmer twin.

By the time that Voyager passed over Neptune's north pole, we knew that the mission had been a success—but there was more to come. Five hours later came the pass of Triton, which, to many of us, ranked as the highlight of the whole encounter.

Triton's diameter had never been properly measured. Generally it was assumed to be about the same size as our Moon, but one estimate made it larger than Mercury! It was known to have an atmosphere, and it was thought that the main constituent was likely to be methane; there was always the chance that the atmosphere would be cloudy enough to hide the surface completely, as with Titan in Saturn's family.

Not so! First, Triton turned out to be smaller than expected, with a diameter of only 1681 miles, less than that of the Moon though rather greater than that of Pluto. If Triton were smaller than had been thought, it also had to be more reflective, and therefore colder. We now know that the surface temperature is −235 degrees Centigrade, which is a

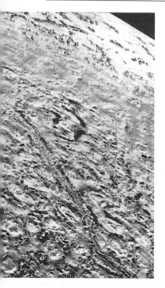

The surface of Triton from Voyager 2: grooved terrain

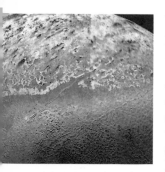

The nitrogen snow cap covering Triton's pole, as imaged from Voyager 2.

mere 38 degrees above absolute zero; it is the chilliest world ever encountered by an Earth-built probe. The density is fairly high (more than twice that of water), and it seems that the composition is about two-thirds rock and one-third ice. Triton is appreciably more substantial than the icy satellites of Saturn and Uranus.

Colour was evident—and it soon became clear that the southern polar region, Uhlanga Regio according to modern nomenclature, is remarkable by any standards. It is in the midst of its immensely long summer, and is covered with what looks like pink snow. But the ice is not water ice; it is nitrogen ice, with an admixture of methane ice. There is almost no surface relief, and there are very few conventional craters, though there are large enclosures which look like frozen lakes—particularly in the northern regions, where the hue is darker and almost bluish; there is a marked demarcation line between the two types of terrain. We need not have worried about clouds. The density of the atmosphere is only about 10 microbars, or 1/100,000 that of our air at sea-level—enough to produce detectable haze over Triton's limb, but nothing else. The main constituent is not methane, but nitrogen.

Perhaps the most surprising discovery of all was that of nitrogen geysers. When these were first suggested, by Laurence Soderblom of the NASA geological team, he admitted that the idea sounded 'crazy', but he added that crazy ideas sometimes turn out to be correct—and this was one of them. Below the surface of Triton there seems to be a region where the pressure is great enough to turn nitrogen into a liquid. If any of this nitrogen 'sea' percolates upward, it will reach an area, only just below the surface, where the pressure is so low that the nitrogen can no longer remain liquid; it bursts forth at high velocity in a shower of nitrogen ice and gas, producing a geyser. The expelled materials are then blown 'downwind' in the tenuous atmosphere to produce dark streaks. Several of these are to be seen in the pink snow-cap of Uhlanga Regio, and there is every reason to believe that the geysers are active at the present time.

What we would really like to do is to have another close look at Triton in a few years' time; there will be changes due to the differing seasonal temperatures, because the nitrogen-cum-methane ice is mobile. Sadly, this is just what we cannot do. Voyager 2 has gone on its way; it sent back one last view of the twin crescents of Neptune and Triton, but within a day or two it had moved so far out that all detail had been lost. No further Neptune probes are funded as yet, and so it may be many decades before we learn more.

We ended our last transmission at the board outside the Deep Space Network, where the position of Voyager had been given daily throughout the encounter; now it had been replaced with a poignant message— *Bon Voyage, Voyager.* Though its work is not done, and we should keep in touch with it until well into the Twenty-first Century, it can send us no more pictures.

So farewell, Voyager 2, and good luck. Will any alien civilization ever find you? I would like to think so!

13 THE UNEXPECTED IN ASTRONOMY

Life is full of surprises—and this is as true in astronomy as it is in every other sphere of existence. The skies are always changing, and one never quite knows what is likely to happen next. This is so even when one sets out to make a purely routine observation, as I did in mid-1989 when the planet Jupiter came round from 'behind' the Sun. Before conjunction it had looked quite normal, with two broad belts, the North Equatorial and the South Equatorial, one to either side of the equator; the famous Great Red Spot had virtually disappeared, as it periodically does (though it always comes back).

I opened the dome of my 15-inch reflector, swung the telescope, and Jupiter swam into the field. For a moment I thought that there was something wrong with the optics—because the South Equatorial Belt had disappeared. I could see no trace of it. Having satisfied myself that there really was nothing wrong, I sent a message through to the International Astronomical Union telegram headquarters in America. In fact it turned out that I was not the first to see this strange appearance—it had been noted a couple of nights before—but at least I was one of its earliest observers.

Then, on 13 January 1990, I made another observation which surprised me: the equatorial region of the planet had turned chrome yellow, and this time I really was the first to report it, though it did not last for long. Meanwhile the Great Red Spot had returned, with the ghost of the South Equatorial Belt.

Jupiter has a gaseous surface, so it is natural for it to change; the Red Spot is now known to be a vast whirling storm—a phenomenon of Jovian 'weather'. Major alterations can be quite rapid, but I have never seen anything quite so dramatic as the disappearance of the South Equatorial Belt. It returned, of course, but not for some time did Jupiter again look really 'normal'.

Saturn, the second giant planet, is less active than Jupiter, probably because it is less massive and is further from the Sun, but it too can 'do the unexpected' as happened in 1933 when it was under scrutiny from a very unusual astronomer—W.T. Hay, whom you may remember better as Will Hay, the stage and screen comedian. Will Hay was a very serious and skilled amateur astronomer, and with his 6-inch refractor he suddenly noticed a brilliant white spot on Saturn's disk. At once he

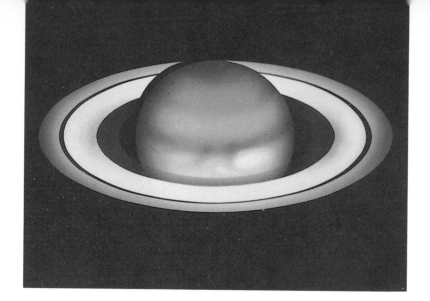

The white spot on Saturn, drawn by Paul Doherty on 14 October 1990 with a 12-in refractor.

Halley's Comet, 1910, when it was much brighter than at the return of 1986.

contacted a colleague, Dr W.H. Steavenson, who confirmed it, and the news spread quickly.

Again it did not last for long; over the following weeks it elongated and became diffuse, until it was no more than a bright region extending round the planet. While it lasted, it was useful in measuring Saturn's rotation period, which was not then known at all accurately. It was some sort of outbreak from below the surface, and nothing quite so striking was seen again until the latest white spot, that of 1990. So next time you see a Will Hay film, respect him as an astronomer as well as a comic schoolmaster!

Comets, of course, are always apt to catch us by surprise. In 1910, just before the predicted return of Halley's Comet, a group of diamond miners in South Africa made an unexpected discovery. Emerging from their mine after a shift, they saw a brilliant comet. It was not Halley's;

it was a much brighter comet which nobody had predicted. It became visible even when the Sun was above the horizon, so that it is generally referred to as the Daylight Comet. There can be no one discoverer after whom it can be named; credit goes to all the workers at the Transvaal Premier Diamond Mine.

By late January the tail had attained a length of 30 degrees, and the comet far outshone Halley's, which was already visible telescopically but did not reach its peak until later. The Daylight Comet faded quickly, but was followed until July 15, when Max Wolf at Heidelberg saw it for the last time as a dim speck of magnitude 16½. It was then well out in the asteroid belt. It will be back—but not for about 4,000,000 years.

When Halley returned again in 1986, I met many elderly people who told me proudly that they had seen it in 1910. Generally I did not have the heart to tell them that in all probability they had seen the Daylight Comet instead!

In 1882 there was an eclipse of the Sun, visible from Egypt. A party of astronomers went there to observe it, and were astonished to see a bright comet, which came into view as soon as the last segment of the Sun was hidden by the advancing Moon and the sky darkened. The comet had never been seen before, and it was never seen again, so that this is our only record of it. We know it as Tewfik's Comet, in honour of the then ruler of Egypt.

Something of the same kind happened on 1 November 1948, when another 'eclipse comet' was found unexpectedly less than two degrees from the centre of the hidden Sun. However, it was subsequently seen by many observers, and was followed until the middle of the following February. It was an easy naked-eye object with a 5-degree tail, though by no stretch of the imagination could it be called 'great'. The period has been given as 95,000 years.

Yet another brilliant comet, that of 1882, caused Sir David Gill, Director of the Cape Observatory, to make a bold decision. Photography in those days was still decidedly primitive, but Gill decided to attempt a picture of the comet. With the aid of a photographer friend he rigged up some equipment, mounted it on his telescope, and produced an excellent print which showed not only the comet, but also an amazing number of stars. Instantly Gill realized that this was the way to map the starry sky—and so that one picture led on directly to the great photographic star atlases of today.

Still on comets, we must remember Biela's, which used to have a period of 6¾ years. Late in 1845 it astounded astronomers by breaking in half. The twins came back on schedule in 1852, were missed in 1858 because they were badly placed, and then vanished. They were expected in 1866, but failed to appear, and have never been seen again, though in 1872 a brilliant meteor shower radiated from that part of the sky where Biela's Comet ought to have been. This was one of the earliest direct proofs of the connection between comets and meteors, which are now known to be cometary débris. The Bielid shower was repeated in 1885, but became weaker in later years, and now seems to be more or less defunct.

On 1 September 1975, I went out to my observatory to make some routine observations of variable stars. One of these was a curious little star in the Swan, SS Cygni, which is normally of about magnitude 12—much too dim to be seen with the naked eye, or even binoculars—but which periodically flares up to above magnitude 9. It is made up of an ordinary dwarf star together with an old, bankrupt White Dwarf. The White Dwarf pulls material away from its more rarefied companion, and this material builds up in a sort of ring. More or less regularly—about every 48 days in the case of SS Cygni—things become unstable, and there is a mild outburst which lasts for a few nights.

On this occasion I saw something else: a bright new star not far from SS. It certainly had not been there on the previous night. I knew at once what it was: a nova, again made up of a normal star together with a White Dwarf, but with the difference that the build-up of material round the White Dwarf is so much greater that eventually there is a really violent explosion. Some past novæ have been striking. This one rivalled the Pole Star, and it completely altered the look of the whole of that part of the sky.

I guessed that I was not the first to see it, but it was just as well to make sure, so I rang the Royal Greenwich Observatory, then at Herstmonceux Castle in Sussex. As I had expected, the Japanese had beaten me to it. Darkness had come earlier to Japan, and observers there had found the star at about 3 p.m. British time. Indeed, I was not the first to see it even from England; I think I was about 83rd on the list of discoverers. All the same, it was an exciting moment.

Karl Jansky's 'Merry-go-round', with which he discovered radio waves from the Milky Way.

The nova soon faded, and within a few weeks had become so dim that I lost it even with my largest telescope. It was, in fact, the 'quickest' nova on record.

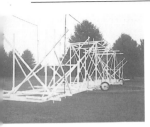

Replica of Jansky's original radio telescope, on display at Green Ban, West Virginia; I photographed it there in 1992.

Nowadays radio astronomy is a vitally important branch of science, but it began in 1931 in a curious manner. Karl Jansky, an American radio engineer of Czech descent, was investigating 'static', using an improvised aerial, when he found that he was picking up long-wavelength radiations from the Milky Way. This was the great 'breakthrough', but, surprisingly, Jansky never followed it up as he might have been expected to do. He published a few papers, which caused no excitement whatsoever, and then dropped the whole subject. Not until after the end of the war did radio astronomy make any real progress.

The region of the electromagnetic spectrum between visible light and radio waves includes the infra-red, which cannot be well studied from ground level because so many of the radiations are blocked by layers in the upper atmosphere. In 1983 the Infra-Red Astronomical Satellite, IRAS, was launched, and operated for most of the year. The signals were collected and analysed at the Rutherford–Appleton Laboratory, not far from Oxford. There, Drs Hartmut ('George') Aumann and Fred Gillett were testing the equipment on board IRAS when they found that the brilliant star Vega had 'a huge infra-red excess', possibly indicating cool, planet-forming material. Nothing of the sort had been expected. They found the same excesses with other stars, notably Fomalhaut in Piscis Australis and Beta Pictoris in the little southern constellation of the Painter; in the case of Beta Pictoris, the material has now been photographed in visible light. Does it indicate that we are looking at another Solar System? It may do; we cannot be sure, but the possibility is there, and it all goes back to that unexpected discovery of cool material surrounding Vega.

Finally, what about phenomena that you cannot possibly miss? I well remember that on the evening of 26 January, 1938, I went out after dark and saw the whole sky glowing red. It took me aback; I realized that it was a display of aurora, but I was totally unprepared for it. At that time I lived at East Grinstead in Sussex, and auroræ are not common in latitudes as far south as that. They are caused by events on the Sun; active regions send out streams of electrified particles which cross the 93,000,000-mile gap between the Sun and the Earth and cascade down into the upper air, making it glow. Because the particles are charged, they tend to spiral toward the magnetic poles; the geographical poles have nothing directly to do with it.

We cannot forecast auroræ widespread enough to be seen all over England, but they do happen sometimes. For instance, there was a grand display on 13 March 1989, which took everyone unawares. From parts of the Midlands it was brilliant enough to cast shadows. There was a large, active sunspot on view at that time, but by no means every active spot produces an aurora.

These are just a few of the surprises which astronomers encounter. It all adds to the fascination of the sky—and remember, we may at any time have another outbreak on Saturn, a nova, an aurora or a brilliant, blazing comet.

14 WHEN IS A STAR NOT A STAR?

Over the past few years a good deal has been heard about Brown Dwarfs—failed stars, which have never become hot enough to shine in the same way as normal stars such as the Sun, and which may be in the nature of missing links between stars and planets. Various claims have been made, but now, for the first time, a Brown Dwarf may have been positively identified. It is known provisionally as MH18; it is about 68 light-years away, and has no more than 1/20,000 the luminosity of the Sun, with only 5 per cent of the solar mass. In this case, it is some 30 per cent dimmer than the previously known faintest star (LHS 2397a), and it does seem to be a Brown Dwarf.

Previous searches for these 'missing links' have involved studying visible stars and looking for dim companions. For example, in 1988 E.E. Becklin and B. Zuckerman located an object near the White Dwarf star GD 165 which seems to be about 120 astronomical units from its

Rich star-field in the Milky Way.

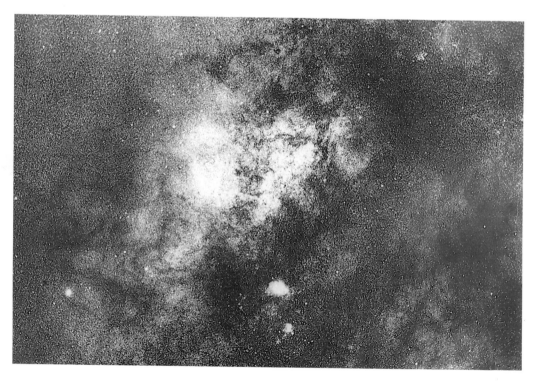

The United Kingdom Schmidt telescope (UKS) at Siding Spring: my photograph, 1984.

companion, with a surface temperature of no more than 2100 degrees and a mass from 6 to 8 per cent that of the Sun, putting it firmly in the Brown Dwarf range; but final proof has been lacking. This is also true of studies by Bruce Campbell and David Latham of the changing toward-or-away velocities of some visible stars which could be due to low-mass companions. But the case of MH18 is rather different.

The story began with some researches by Dr Mike Hawkins, of the Royal Observatory Edinburgh. He was using Schmidt plates, taken with the UK Schmidt telescope at Siding Spring in New South Wales, to search for cool White Dwarfs. A White Dwarf is a very old star which has used up its main store of nuclear energy and has collapsed, so that it consists only of a very small, hot, dense core with no energy reserves left. The density is tremendous—over 60,000 times that of water in the case of the Companion of Sirius, which is by no means the most extreme example known. But as a White Dwarf is shining only because it is still slowly shrinking, it ought to cool down, and all known White Dwarfs are very hot. Why are there no cool specimens? It could be that the universe is not yet old enough to have allowed a White Dwarf to cool, and that is what Hawkins wanted to find out.

He did not find any cool White Dwarfs, but he did locate some stars which appeared to be very dim and very red, so that they showed up best on infra-red plates. This was interesting. Unfortunately, some of the most promising candidates were single stars, and determining the mass of a single star is notoriously difficult. The first step is to measure the distance, which has to be done by the method of parallax—essentially the same as the method used by a surveyor to find the distance of some inaccessible object, such as a mountain-top; with a star, the position against the background of more remote stars is measured from opposite ends of the Earth's orbit, giving a baseline of 186,000,000 miles. When this was done with MH18, the result came as quite a shock. The star

A Brown Dwarf. This is an impression by Paul Doherty, but is probably very like the actual scene of a Brown Dwarf—assuming that such stars really exist!

was much closer than had been expected, because on the cosmical scale 68 light-years is not far. It followed that the star also had to be much dimmer than had been thought.

Next, the spectrum was examined, and this confirmed that the star really was a dwarf, with a surface temperature of the order of 2000 degrees. All in all, there seems little doubt that the mass is below 8 per cent of that of the Sun, which is the lower limit for a star capable of triggering off nuclear reactions and starting to shine in the usual way.

It is important to stress that there is an essential difference between a White Dwarf, which is the end product of a normal star, and a Brown Dwarf, which is not. Because a Brown Dwarf depends for its dim luminosity upon gravity and nothing else, it does not 'shine' for long—perhaps for no more than a thousand million years—and because of its faintness, it will not be detectable unless it is reasonably close. We know that there is a great deal of 'missing mass' in the Galaxy; because of the way in which the stars move around, we can be sure that there is a tremendous quantity of material which we cannot actually see. It is tempting to suggest that this could be due to swarms of unseen Brown Dwarfs. This could not account for the missing mass in clusters of galaxies, but it could be a factor in the neighbourhood of the Sun.

It is not easy to give a dividing line between a Brown Dwarf and a planet. It may be said that a Brown Dwarf is a body which is (just) hot enough to shine detectably, whereas a planet such as Jupiter is not. The limit may be about ten Jupiter masses, bearing in mind that Jupiter has 318 times the mass of the Earth. However, we do not really know, and we must await the detection and investigation of more Brown Dwarfs.

Seen from close range, a Brown Dwarf would be a strange world. It would be too hot for a landing to be made, and yet it is very cool by stellar standards. As yet the Brown Dwarfs remain mysterious, but at least we are fairly confident that we have made the first reliable identification of one of these missing links—bodies which are not quite stars and not quite planets.

15 THE WILLIAM HERSCHEL TELESCOPE

The island of La Palma, in the Canaries, is not a popular tourist resort in the style of Las Palmas or Tenerife, but it is of special importance to scientists. Atop the extinct volcano of Los Muchachos there is one of the world's greatest observatories. La Palma is a Spanish island, but the observatory is truly international.

When I first went there, the road up the volcano was very rough—if it could be called a 'road' at all; several times my Land Rover was stuck in potholes. Today the road is excellent, and driving up it is no problem at all. From the summit the view is superb; on a clear day you can even see Mount Teide, on Tenerife, which is as far from Los Muchachos as London is from Birmingham. There is an impressive caldera, and there are the 'rocks' after which the Observatory is named—Los Muchachos, the 'Boys'. I am sure that there must be a local legend about them, though I have never been able to find out what it is.

Dome of the INT (Isaac Newton Telescope) on La Palma.

The William Herschel Telescope on La Palma. I am standing beside it, to show its size.

There are various telescopes on the volcano. The first was the Swedish tower telescope, devoted entirely to studies of the Sun. Others include the 100-inch Nordic telescope, a combined project from Norway, Sweden, Denmark and Finland; the Carlsberg automatic transit instrument, and the 1-metre Jacobus Kapteyn telescope, named in honour of the Dutch astronomer who discovered the phenomenon of star-streaming. Another instrument is the German gamma-ray array, which at first sight looks rather like a collection of telephone kiosks. Seeing conditions are excellent, which of course is why the telescopes are here; very often you can see the cloud-base below, looking remarkably like the surface of the sea.

British involvement began early, with the decision to transfer our then largest telescope, the 100-inch Isaac Newton reflector, from Herstmonceux in Sussex to a clearer site.* It has now been in operation at Los Muchachos for years. However, it has been dwarfed by our latest acquisition, the 4.2-metre or 165-inch William Herschel reflector, one of the most powerful ever built.

In appearance it is rather squat. The tube is a skeleton, as with all large reflectors, and the mounting is altazimuth—that is to say there are two separate movements, up and down (altitude) and east to west (azimuth). This type of mounting means that the cost of the telescope, with its dome, is less than with the old equatorial type. The drawback

---

*Subsequently the Science and Engineering Research Council, which controls all the finances, closed Herstmonceux completely and transferred the rump of the Royal Greenwich Observatory to an office block in Cambridge. This was against the advice of practically every astronomer in the country. I became involved in the whole episode, because I was asked to do so; I have told part of the story elsewhere (*Fireside Astronomy*, John Wiley & Sons, 1992, Chapter 35), but much more could be said.

is that the telescope does not naturally track the stars as they shift across the sky by virtue of the Earth's rotation. However, modern computers have no trouble in dealing with this. The track information is updated twenty times a second, so that the accuracy is at least as good as one second of arc.

The William Herschel telescope has five focal stations, where instruments can be put. Two of these are the Nasmyth foci—strong platforms, capable of bearing very heavy equipment. On one of these platforms is a thermally controlled hut housing instruments dealing with high-resolution imaging, while on the other there is a Dutch-built spectrograph used for investigating the spectra of remote objects such as quasars.

Nobody ever looks through the William Herschel Telescope. Where the eyepiece might be expected to be, we find ISIS, a double-beam conventional spectrograph. As light passes into ISIS the beam is split into two parts, blue and red, each of which then goes to a diffraction grating, which looks rather like a plane mirror, but has rulings which split up the light into its various wavelengths. Even the coarsest grating in use has 158 rulings per millimetre!

Another instrument which can be used in this position on the William Herschel Telescope is TAURUS, used to investigate shells round elliptical galaxies. These shells are produced when two galaxies merge— quite recently on the cosmical scale; perhaps only ten to a hundred million years ago. When a small galaxy falls into an elliptical system, sharp-edged structures are produced, and can tell us a great deal not only about the galaxies concerned but also about the distribution of mass in the parent galaxy. The distribution of mass is often very different from the distribution of light sources, and this is one way of investigating the so-called 'dark matter' which may well make up a very large fraction of the total mass of the universe.

During my visit in the summer of 1990 I was able to talk with Professor John Baldwin, who was using the William Herschel Telescope to record the surface details on a star. This had never previously been done with any success, and is clearly immensely difficult. The star with the largest apparent diameter (not counting the Sun, naturally) is Betelgeux, but even here the apparent diameter is a mere 50 milliarc seconds, which is about the same as that of a three-foot ruler placed on the surface of the Moon and observed from Earth.

Betelgeux was the subject of John Baldwin's investigation. It is a huge red supergiant, over 15,000 times as luminous as the Sun and several hundreds of light-years away (the official Cambridge catalogue gives its distance as 310 light-years, but there have been suggestions that this may be an underestimate). Certainly it is big enough to contain the whole path of the Earth round the Sun, but it still appears virtually as a point source.

It had been proposed that the outer layer of a Betelgeux-type supergiant should consist of a few vast convection cells, where energy bubbles up from below. Using complicated electronic methods, Baldwin and his team found that this is correct, and there was one bright patch near the edge of the tiny disk which gave every indication of being a convection cell.

The Faint Object Spectrograph, part of ISIS, is being used to measure the spectra of very dim galaxies, mainly those which had been observed at infra-red wavelengths by IRAS (the Infra-Red Astronomical Satellite) during its brief but profitable career in 1983. The amount by which the light from a galaxy is shifted to the red or long-wave end of the spectral band shows how fast it is moving away from us. The spectrograph on the William Herschel Telescope has now provided red shifts for about 4000 of these special galaxies, and has led on to some very interesting cosmological studies. For example, there seem to be surprisingly strong changes in the galaxy population; only a few thousand million years ago, galaxies were undergoing frequent bursts of star formation, possibly because collisions between galaxies were more common then than they are now.

The great light-grasp of the telescope, and its rapid speed of operation, means that it is ideal for studying very faint objects. For example: when Edwin Hubble discovered the expansion of the universe, using the 100-inch Hooker reflector at Mount Wilson during the 1920s, he could do no better than measure one red shift per night. On the William Herschel Telescope, it is not hard to measure well over a hundred red shifts in one night's observing.

Of course, these are only a few of the many programmes being pursued with the telescope, which is in action on every clear night. Much has been accomplished already, and much more will be done in the near future. We can all be proud of the William Herschel Telescope.

16 DAYLIGHT STAR

Though this book is called *The Sky at Night*, we must not exclude the sky by day—because the nearest star is the Sun. Moreover the Sun is a perfectly normal, run-of-the-mill star, much less powerful than many of the 100,000 million others in our Galaxy.

By conventional standards it is huge. Its body could contain over a million globes the volume of the Earth, but it is not a million times as massive, because it is not solid; the surface is made up of very hot gas, at a temperature not far short of 6000 degrees Centigrade. Near the core, where the energy is being produced, the temperature rises to at least 14,000,000 degrees, and possibly rather more. The Sun shines by nuclear reactions, chiefly by converting hydrogen into helium. The loss of mass amounts to 4,000,000 tons every second, but by solar standards this is negligible, and there will be no obvious alteration for several thousands of millions of years yet.

To look directly at the Sun through any telescope, even with the use of a dark filter, is very dangerous indeed, and will result in permanent damage to the eye. The only safe method is that of projection, first used by Christoph Scheiner not long after the development of the earliest telescopes of which we have any positive knowledge—around 1610. Simply attach a projection screen to the telescope to hold a card on which the solar image will appear.

One advantage of the projection method is that sunspot positions can be marked directly on the card. A 6-inch circle is drawn in advance, and the Sun's image fitted into it. The circle should have a number of east-west parallel lines drawn on it, and if a spot is seen the Earth's rotation should be allowed to make the spot drift along one of these lines; by turning the projection screen, the correct orientation can be found. Once a spot's position has been fixed, its solar latitude and longitude can be worked out. (Of course, there is no need to use a 6-inch circle; it is simply a matter of convenience—and if you do not want to go to the trouble of making a permanent attachment, there is no harm in holding the card by hand, though obviously the results will be less precise.)

Any type of telescope can be used, but refractors are preferable to reflectors; with a reflector, the mirror should be stopped down to two or three inches, as otherwise the shapes of the optics may be distorted by the Sun's heat. A Schmidt-Cassegrain telescope should also be stopped down, because excess heat can damage the adhesive which holds the secondary mirror on to its mounting.

Projecting the Sun: This is the McClure telescope at the Norman Lockyer Observatory, at Sidmouth in Devon.

I have often been accused of being alarmist in my persistent stressing of the dangers connected with observing the Sun. This may be true, but I make no apologies, probably because of an experience I had when I was about ten years old. I met an elderly amateur astronomer who was blind in one eye—and had been so ever since he was a teenager; he had been looking directly at the Sun with a 3-inch refractor, using a dark suncap over the eyepiece, when the filter shattered so quickly that he had no time to move his eye out of the way. That taught me a lesson, and I say again that unless you are a genuine expert there is only one rule about looking straight at the Sun through any telescope: *Don't.*

Sunspots are formed where magnetic fields inside the Sun break through the surface, cooling it down. There is a semi-regular eleven-year cycle, which has been known ever since 1843 as a result of long-continued observations by a German amateur, Heinrich Schwabe (though in fact Schwabe was not particularly concerned with the Sun itself; he was hoping to catch sight of an unknown inner planet crossing in transit over the solar disk). Rudolf Wolf, Director of the Berne Observatory and later of Zürich, collected as many observations as he could going back to 1610, and was able to fix the dates of maximum and minimum activity between 1710 and 1850 with reasonable accuracy. He found that the average period of the cycle was 11.1 years, but neither the period nor the strength of maximum is constant; thus the maximum of 1957–8 was about four times as energetic as that of 1816, which was particularly weak. The present cycle (No. 22) is of the 'high' type, and considerably more active than Cycle 21, which peaked in 1980.

The situation may not always have been the same. So far as we can judge from the incomplete records, there was a prolonged minimum between 1645 and 1715 when there were almost no spots at all; this is usually called the Maunder Minimum, because attention was drawn to it by E.W. Maunder almost a century ago now. Perhaps significantly, this was also a very cold period. Auroræ were almost never seen, and when the Sun was totally eclipsed there was not much in the way of a corona. Just why this happened is not known—it seems almost as though the Sun 'switched off' for seventy years.

Rudolf Wolf worked out a method of giving monthly averages of activity. Each group counts 10 points, and each individual umbra as 1 (the umbra being the darkest, central part of a spot). The Wolf or Zürich number for any day may range between zero and 300. The latest maximum fell at the end of 1990, so that during the next few years we may expect many days when there are no groups at all.

The latitudes of spots on the disk are regulated by what is termed Spörer's Law—named after the German astronomer Gustav Spörer, though in fact the same sort of thing had been noted rather earlier by an enthusiastic British amateur, Richard Carrington. At the start of a new cycle, spots break out at around 30 degrees north or south. As the cycle progresses, spots occur nearer and nearer the equator, until they reach around latitude 15 degrees at the peak of the cycle and then tending to cluster roughly 5 degrees to either side of the equator as activity dies away. At minimum, the first spots of the new cycle may start to appear before the last groups of the old cycle have vanished. Note that individual spots do not migrate toward the equator; it is only the latitudes in which new spots are likely to form which actually shift. Moreover, not even a large spot-group can persist for more than a few months, and small spots may have lifetimes of only a few hours.

Carrington's observations of 1859 also showed that the Sun does not rotate in the way that a rigid body would do. The rotation period at the equator is 25 days, but at the poles it is nearer 35 days. This so-called 'differential rotation' is of great importance when we try to work out just how the Sun functions.

As yet there is no complete explanation of the solar cycle, but it is certainly an essentially magnetic phenomenon, and for many years now we have known that sunspots are associated with powerful magnetic fields. Tracking them from day to day is a fascinating pastime, and the changes are obvious even over short periods.

In particular, the so-called Wilson effect is worth checking. Consider a regular spot, with a dark central umbra and a lighter surrounding penumbra. When near the centre of the disk, it will appear symmetrical; but as it nears the edge and becomes foreshortened, the penumbra will

Alexander Wilson (1714–1786), the Scottish scientist who first described the 'Wilson Effect' in sunspots.

always appear broadened toward the limbward side. In 1769 the Scottish astronomer Alexander Wilson noted this, and suggested that the spots must be depressions rather than humps. Not all spots show the effect, and of course many groups are very irregular, with many umbræ enclosed in a penumbral mass, but even here the effect can often be detected.

In white light, the normal surface of the Sun (known as the photosphere) has a granular appearance, marking the top of the turbulent zone where convection carries energy upward from the Sun's interior. Indeed, the surface is 'boiling' all the time. Each granule lasts for from five to ten minutes, and has a diameter of between 450 and 650 miles; the velocities may reach several miles per second. Time-lapse photography with equipment such as the Swedish solar telescope on Los Muchachos also makes it possible to see the way in which material flows in and out of the various spots and groups.

With amateur equipment, granulation can be observed in white light only under excellent seeing conditions, but many amateurs now have access to narrow-band hydrogen-alpha filters, which transmit only a very narrow wavelength of the electromagnetic spectrum at 6563 Ångströms. By isolating the light of hydrogen, such filters make it possible to see solar prominences—great loops of relatively cool hydrogen standing or shooting above the Sun's surface, often reaching out to more than 30,000 miles. One prominence, seen on 23 April 1989, extended up to almost 200,000 miles.

Prominences may be either active or quiescent. At the limb of the Sun they stand out as flame-like structures (hence the old, misleading name of Red Flames), but against the disk they show up as dark features, and are then called filaments. They are not completely dark, but are less brilliant than the photosphere, and seem black when viewed against the disk in H-alpha. With the naked eye they can be seen only during a total eclipse, when the whole of the Sun's bright body is blotted out by the Moon.

Narrow-band filters can also be used to study flares, which are the most energetic and spectacular of all solar phenomena, and are responsible for magnetic storms on Earth as well as displays of auroræ. The magnetic fields of sunspots store vast amounts of energy, some of which may be released suddenly in a flare; when the field lines reconnect, large currents flow, and in a few minutes the temperature rises to many millions of degrees. The major spot-group of March 1989, which was responsible for the great aurora of March 13, shot out 195 flares during its passage across the visible disk. The largest, at 19.22 hours GMT on March 10, released energy equivalent to 10,000 million megaton bombs.

The Sun's outer atmosphere, the corona, is also visible with the naked eye during a total eclipse, but is difficult to study at any other time simply because it is so faint. It too is affected by the solar cycle. Near maximum it is extensive all round the Sun, with tapering streamers stretching outward like the petals of a flower, whereas at minimum the

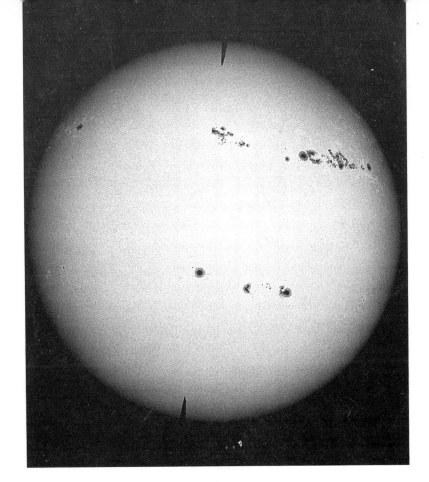

Active solar disk: 20 February 1956. Photograph taken at Mount Wilson (California). This was not far from solar maximum.

corona shrinks, with streamers more or less confined to the plane of the Sun's equator.

Since the Earth's weather system is driven mainly by solar energy, it is reasonable to suppose that even a slight variation in solar output could affect our weather and climate. This may well be true, but the situation is still unclear.

It has long been assumed that the Sun's output of radiation is constant, apart from small variations due to spots. Spots may cover 1 to 2 per cent of the disk at times, and since they are at a rather lower temperature than the rest of the surface they would be expected to cause a slight reduction in total output—perhaps by 0.1 to 0.2 per cent at solar maximum. In 1979 it became possible to measure such small variations directly, using space-borne equipment, and to the general surprise it was found that the output is highest not at minimum, but at solar maximum. As the numbers of spots increase, the output rises. Apparently this is because major groups are usually surrounded by faculæ, which are very brilliant and compensate for the blocking effects of the spots themselves.

Over longer time-scales, the evidence for a link between solar activity and terrestrial climate seems to be reasonably strong. The main clue comes from the behaviour of cosmic rays, which are not rays at all, but high speed atomic particles coming in from all directions all the time.

Total eclipse of the Sun, 11 July 1991, as I photographed it from Mexico.

At solar maximum, magnetic effects due to the Sun deflect some of these cosmic rays away from the Earth, and this in turn has an effect upon the amount of a substance called carbon-14 which is found in the atmosphere. When trees assimilate carbon dioxide, each growth ring contains a small amount of carbon-14, and it follows that this amount is least when the Sun is most active, so that tree-ring studies can tell us a great deal about the way in which the Sun has fluctuated. Incidentally, the pioneer worker in the field of tree-ring research was an astronomer: A.E. Douglass, who spent many years at the Lowell Observatory in Arizona.

In 1976 Jack Eddy compared the carbon-14 record of solar activity with historical sightings of sunspots and auroræ, as well as climatic data. He confirmed the Maunder Minimum of 1645-1715, which, as we have seen, was also marked by a cold period in Europe known as the Little Ice Age, with temperatures about one degree below normal. All of which is highly interesting, but the evidence is not conclusive, and so far we have to admit that we are still not sure whether there is any close connection between sunspots and weather.

In any case, we know that so far as we are concerned the Sun is all-important. It may be a very average star in a very average Galaxy; but without it, you and I would not exist.

17 MISSION TO TITAN

Of all the worlds in the Solar System, one of the most intriguing is Titan, the largest member of Saturn's family of satellites. It was discovered by the Dutch observer Christiaan Huygens as long ago as 1655, and is bright enough to be visible with a small telescope; it has been said that really keen-eyed people have been able to glimpse it with binoculars (though I have never been able to do so myself). It is big, with a diameter of 3201 miles, so that it is actually larger than the planet Mercury, though less massive. Of all planetary satellites, only Ganymede in Jupiter's system has a greater diameter—and then only by 73 miles.

Titan moves around Saturn at a mean distance of 759,350 miles in a period of 16 days. Its escape velocity is 1.54 miles per second, which is very slightly greater than that of our Moon (1.48 miles per second). Yet the Moon is airless, while as early as 1944 it was discovered—spectroscopically, by Gerard Kuiper—that Titan has an appreciable atmosphere. Initially it was assumed that the atmosphere must be tenuous, and probably made up chiefly of methane, but the Voyager 1 pass of 1980 showed that both these assumptions were wrong. The atmosphere is dense, and is made up mainly of nitrogen, though there is a good deal of methane as well, together with a little free oxygen. We can never see through the clouds of Titan; all we can make out

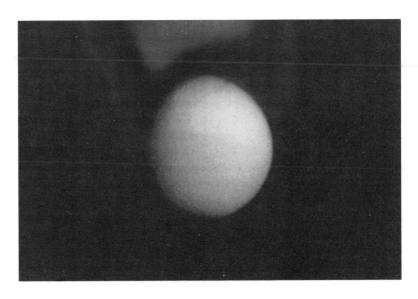

Titan, from 2,800,000 miles: Voyager 1. The actual surface is hidden by the dense, cloud-laden atmosphere.

Impression of the Cassini probe passing Saturn.

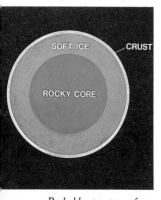

Probable structure of Titan.

is the top of a layer of orange smog. All the ingredients for life exist there, though the low temperature seems to indicate that life has never been able to appear.

Most of our detailed knowledge of Titan comes from Voyager 1. Had its twin, Voyager 2, been routed to image Titan there would have been no chance of going on to Uranus and Neptune, because the probe's path would have been sent well out of the ecliptic, and in any case Voyager 2 could have added little to what Voyager 1 had already told us. So a new mission is needed, and one has been planned: Cassini, scheduled for launch in April 1996. This provides a ten-day 'launch window', and is the best chance for the next twenty years. The launching vehicle will be an expendable Atlas/Centaur.

Let us look first at Titan's atmosphere. The surface pressure is about 60 per cent higher than here on Earth, but this fails to give the full impression of its extent. Were you to hold out your hand, you could in principle measure the quantity or mass of gas above your hand stretching up to the very top of the atmosphere. If you were to do exactly the same experiment on Titan, you would find that the quantity of gas would be ten times as great.

Nitrogen makes up 90 per cent of the Titanian atmosphere. Methane accounts for most of the rest, with less than 1 per cent of hydrogen molecules. However, a dozen or more hydrocarbon gases were also detected at lower concentrations by Voyager 1. These are of great interest, and show that the atmosphere is a rich laboratory for a whole variety of complex chemical reactions which may well have been triggered off by the action of sunlight. It has even been suggested that some of these reactions are similar to those which occurred on Earth as part of a chain

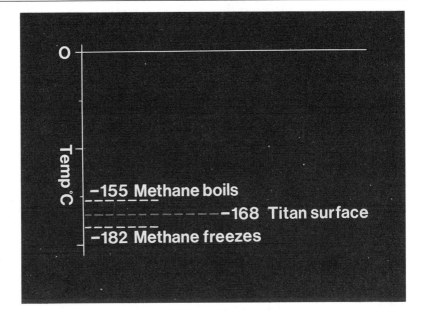

Freezing and melting points of methane. Titan's surface temperature means that methane can therefore exist as a solid, liquid or gas.

which led eventually to the development of life. If, then, we can understand Titan's chemistry, we may have a better understanding of the early history of our own world. But there is always the problem of the low temperature: −168 degrees Centigrade at the surface, rising to −115 degrees Centigrade, in the upper atmosphere. The temperature would be even lower if Titan did not have its own mini-greenhouse effect.

The Cassini probe—named after G.D. Cassini, the seventeenth-century Italian astronomer who discovered four of Saturn's satellites as well as the main gap in the ring-system—will have an unusual trajectory. We have to use the 'slingshot' principle twice in order to provide enough energy to reach Saturn. A fly-by of the Earth itself one year after launch will provide enough energy to move out to the orbit of Jupiter, and once there another gravity-assist manœuvre will provide enough speed to reach Saturn. This route means that the whole journey to Saturn will take 6 years, and there will be two passes through the asteroid belt. On at least one of these there will be a good chance of imaging an asteroid: No. 66 Maja, which was discovered by Horace Tuttle in 1861 and has a diameter of 53 miles.

The proposed sequence of events is as follows:

Launch: April 1996.
Asteroid fly-by (Maja): 1997.
Earth fly-by: 1998.
Second passage through the asteroid belt: 1999.
Jupiter fly-by: 2000.
Arrival at Saturn: 2002.

Both Saturn and Titan are prime targets. This is achieved by having two separate vehicles. One is the orbiter, provided by NASA; the other is the Titan lander, supplied by the European Space Agency (ESA). They will travel together for most of the journey, but will have very different objectives when they reach their targets.

As soon as the complete vehicle arrives, it will be put into a closed path round Saturn, with a revolution period of four months. Some three months into the initial orbit, in January 2003, there will be the first encounter with Titan. Twelve days before closest approach, the lander—named Huygens, in honour of Titan's discoverer—will be separated. The orbiter will be deflected so that it misses Titan altogether, and from then on the two parts of the original space-craft will operate quite independently.

The orbiter will begin what is referred to as the 'Saturn tour', involving about sixty complete orbits before contact is lost. The orbit will be continually changed, giving us many different fly-by encounters of Titan and Saturn over a period which may be as long as four years. Moreover, some of the icy satellites will be imaged from close range, and details on them down to a resolution of about 30 feet should be possible, which is much better than anything managed by the Voyagers.

Then, of course, there are the magnificent rings, which will be studied under a variety of different geometries and over an extended time-scale. This is particularly important, because it is now thought that some of the ring phenomena, such as the puzzling 'spokes' in Ring B, may be short-lived. There will also be an opportunity to examine the Saturnian magnetosphere (that is to say, the region round the planet in which the magnetic field interacts with the solar wind). Altogether the orbiter will carry fifteen instruments, including a camera, a radar mapper, spectrometers extending from the near ultra-violet through infra-red and into the microwave region, and instruments to measure the fields and particles in the magnetosphere. There will also be a dust detector.

Quite apart from all this, the orbiter has another vital rôle to play: it will have to relay the data sent back by the Huygens lander. As we have seen, Huygens will be separated twelve days before arrival at Titan. It will then be set up for its final approach, and will eventually hit the top of Titan's atmosphere at the tremendous speed of about 3.7 miles per second. This is where its deliberately unaerodynamic shape will be essential. The large, dish-shaped structure will be strongly decelerated as it hits the atmosphere, and will be slowed down to a mere 1300 feet per second, at which point the large dish part of the probe can be jettisoned and the parachute released, slowing Huygens down still further and allowing it to float gently. It should take about three hours to reach the surface, and most of the experiments will be switched on during this time, mainly for studies of the atmosphere—including direct sampling of the gases and aerosols making up the clouds. There will also be a camera, a lightning detector, and an instrument designed to study the surface of Titan.

As Huygens disappears beneath the haze and clouds it will be entering regions we have never seen before, and will send back a steady stream of data to the orbiter. Finally it will hit Titan at the relatively sedate speed of just over 3 miles per second. What will it find? Land, or 'sea'?

Water is, of course, out of the question, but it is quite likely that there may be an ocean of ethane or methane. We know that there is a good deal of methane in the atmosphere, but we also know that this methane is broken down by sunlight, and that the process is irreversible; therefore there must be a constant source of replenishment, and this must presumably come direct from Titan's surface. We have also found that the surface temperature is close to the 'triple point' of methane, so that this substance can exist either as a liquid, a solid or a gas (just as H_2O can do on Earth, as water, ice or water vapour). The pressure, too, is suited to a methane ocean. There may even be waves which could be measured by Huygens after landing.

Yet we will have to be quick, because Huygens has a very short life-expectancy after arrival. We are reckoning in terms of a few minutes at most, and possibly no more than 180 seconds. There are various ways in which the probe could die. It could be put out of action by the intense cold; it could simply sink into the ocean, but in any case there is bound to be a loss of the telemetry link, because the orbiter will soon pass out of range, and there is no chance of Huygens lasting until it comes round again. It is rather daunting to reflect that we can hope for less than five minutes' data—probably less—from a space-craft which has taken years to plan and has been on its way for several years more!

We can only hope that the Cassini mission will succeed. Until it arrives, we cannot hope to learn a great deal more about Titan, though ground-based observations may be of at least some use, particularly in the infra-red.

One final point is worth making. We know that in the far future the Sun will become more luminous, finally changing into a Red Giant star. The Earth will be destroyed; the temperature of the outer Solar System will rise, and there have been suggestions that this warming may turn Titan into a habitable place. Alas for such hopes! Increased warmth will mean that the particles in Titan's atmosphere will speed up, and will break free; remember, the actual escape velocity is low.

Although we can hardly hope to find natural life on Titan, either now or in the future, it remains one of the most interesting of all worlds. Land or sea? We do not yet know. With luck, Huygens will tell us.

Added in proof: There are now suggestions that the launch may be delayed by two years. However, the sequence of events will be the same.

18 THE UNVEILING OF VENUS

There was a time, not so long ago, when Venus was regarded as 'the planet of mystery'. From Earth we can do no more than study its upper clouds, and before the Space Age our ignorance of the surface conditions was virtually complete. Venus might be covered with ocean, or it might be an arid, scorching-hot dust-desert; we simply did not know.

The earliest probes, launched during the 1960s, put paid to the attractive marine theory, and subsequent controlled landings by Russian space-craft showed that Venus is a most unfriendly place. The surface temperature is not far short of 1000 degrees Fahrenheit, the atmospheric pressure is at least 90 times greater than that of the Earth's air at sea-level, the atmosphere is made up mainly of carbon dioxide, and the clouds are rich in sulphuric acid. Any possibility of life there seemed to be ruled out. But now, for the first time, we have been able to obtain detailed maps of the surface. They are due to one very successful space-craft, Magellan.

Magellan was launched on 5 May 1989, and reached the neighbourhood of Venus in August of the following year. It was not designed to land, and in fact it had only one objective: to map Venus by radar. It is moving in a closed orbit round the planet, with a minimum distance of about 180 miles and a period of 3.17 hours; its path takes it more or less over Venus' poles. It can collect images for only about forty minutes per orbit, but it needs all the remaining time to transmit the data back to Earth. There have been various alarms—not long after it began its main programme, all contact with it was lost for more than twelve hours—but on the whole it has worked extremely well.

Magellan has three main antennæ. One of these sends a radar pulse vertically downward, and can determine the altitudes of features directly below to an accuracy of about a hundred feet. The 12-foot dish known as the HGA or High-Gain Antenna 'illuminates' the surface below, just as sunlight does, and the surface rocks modify the pulse according to the known laws of physics before it is reflected back to the HGA. We then have to interpret the returned signal. Basically, the amount of scattering from the surface gives a clue to the roughness of the landscape, and the strength of the returned pulse tells us the type of surface material present. There is also a third antenna, used to measure the thermal properties of Venus' lower atmosphere.

The Russian space-probe pictures obtained direct from the surface show orange rocks, but in fact the true colour is grey; they simply reflect the colour of the sky—and it is always cloudy on Venus! The rocks are so hot that they are comparatively buoyant, so that it will be difficult for internal movement to pull them down into the planet's interior; there may be no 'subduction zones' of the same type as those of Earth. Weathering and erosional processes will also be quite different from ours.

Upland rolling plains account for 65 per cent of the surface, highlands for 8 per cent, and lowlands for the rest. The higher regions tend to be rougher than the lowlands, and in radar this means that they are brighter; in a radar image, the bright areas indicate roughness. There are two main 'continents', Ishtar Terra and Aphrodite Terra. (The International Astronomical Union has decreed that all names of features on Venus must be female, which leads to some very peculiar designations!) Ishtar, about the size of Australia, lies in the northern hemisphere; Aphrodite is mainly in the south, though it does straddle the equator. The highest mountains rise to over 5 miles above the adjoining plateau in Ishtar Terra.

The Magellan pictures are amazingly detailed. Look, for example, at Sapas Mons, a mile-high volcano with a base almost 250 miles broad. It seems to be a typical shield volcano, similar to those of Hawaii but considerably larger. Near the summit are groups of pits, probably formed when underground chambers of magma were drained through other sub-surface tubes and the whole outer surface collapsed. There are other volcanoes—for example Sif Mons, Gula Mons and Sappho Patera, surrounded by lava flows emerging from their summits; in the same area is the curious feature now named Heng-O, which is larger than the state of California and was probably formed by movements inside Venus. There are curious dome-like hills, averaging 15 miles in diameter and 2500 feet in height, which can be interpreted as very thick lava-flows which came from an opening on relatively level ground. There are highly fractured domes, lava-flows in many places, and extraordinary features—unique to Venus—which have become known as arachnoids, because they resemble spiders and cobwebs. Some of them are over 140 miles across, and seem to be volcanic in origin, though we cannot as yet pretend to give a full explanation of them.

Just beyond the eastern end of Aphrodite is an upland zone, running roughly north–south, which is made up of Beta Regio and Phœbe Regio; it consists of two domes in Venus' crust which are cut across by rift faults, very similar in form and scale to our own East African Rift Valley. Superimposed on all this are two large shield volcanoes, Rhea Mons and Theia Mons, which give every impression of being active. Some images from Magellan show lobate volcanic flows which have flooded the floor of the central rift, and analysis shows that the surface rocks are of basaltic type, which is no surprise at all. Because the volcanoes on Venus are so large, they must have remained over very long-lived 'hot spots' in the mantle, so that the crust of the planet cannot be sliding along in the same way as that of the Earth. In other words, Venus has

Facing page: The Lada Region of Venus: 47° S, 25° E: Magellan, 1991. The area covered measures 350×390 miles. Lava flows are shown, emanating from the crater Ammavaru, which is some 186 miles to the west.

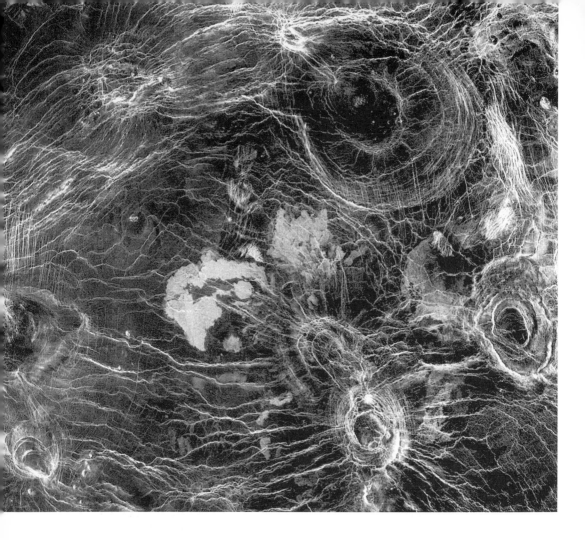

'Arachnoids' on Venus:
Magellan, 1991. These
curious features are
smaller than, but
similar to, the coronæ;
circular, volcanic
structures, associated
with ridges, grooves and
radial streaks. This
image is centred at
latitude 40° N,
longitude 10° E. The
diameters of the
arachnoids range
between about 30 and
140 miles.

no comparable plate tectonics. If it had, then the volcanoes would have stopped growing as soon as their magma supply was moved away from underneath them.

Water plays a major rôle in the plate tectonics of the Earth, and it is this water which is lacking on Venus. It may well be that in the early days of the Solar System, the two worlds began to evolve along the same lines; but as the Sun became more powerful, Venus suffered a runaway greenhouse efect. The oceans boiled away, the carbonates were driven out of the rocks, and in a relatively short time Venus changed from a potentially life-bearing world into the furnace-like environment of today. If life ever gained a foothold there, it will have been unable to survive for very long.

In the late 1950s I remember giving a lecture in Cambridge, dealing with Venus. I said then that it could well be a more friendly place than Mars. I was wrong; Magellan and its predecessors have shown that it is as hostile as it could possibly be. Conditions on the 'Planet of Love' are much more akin to those in the conventional idea of hell.

19 ARMAGH OBSERVATORY: TWO HUNDRED YEARS OLD

Armagh, in Northern Ireland, is not a large city. It lies inland from Belfast, and boasts of two cathedrals, one Protestant and one Catholic. It is also the site of the province's oldest observatory, founded in 1790 as a result of efforts by the current Archbishop, Robinson.

Robinson wanted to establish a university in Armagh. This was never achieved, but the Observatory fitted well into his plans. The first two Directors, Hamilton and Davenport, made little impact (indeed, the luckless Dr Davenport committed suicide in the Director's study, allegedly because his wife was 'an absolute fiend'), but then came Dr Thomas Romney Robinson—no relation to the Archbishop—whose régime extended to over sixty years. He was a colourful character, and a determined one; he equipped the Observatory with new instruments, and even caused a major inter-city railway line to be diverted because he feared that the vibration of passing trains would shake his telescopes!

The Armagh Planetarium.

Dome of the 10-inch Romney Robinson Telescope at Armagh Observatory.

The first of the telescopes was the Troughton Equatorial, delivered in 1795. Robinson used it, together with other instruments, to produce the Armagh Star Catalogue, which came out in 1859 and was a major contribution to positional astronomy. He was succeeded by a Dane, J.L.E. Dreyer, who drew up the 'New General Catalogue' of star-clusters and nebulæ which is still used. There followed a less distinguished period, but then, just before the second world war, Dr Eric Mervyn Lindsay became Director, modernizing and revitalizing the ailing Observatory. He also set up the associated Planetarium, today one of the most famous in the world.

It would be idle to pretend that the Armagh climate is suited to astronomical observation, and today the only notable telescope there is the 10-inch refractor, which is very good indeed (as I know, because I have used it many times). Therefore, Armagh astronomers go abroad to make their observations, and the Observatory is now in the nature of an astronomical institute based in the lovely old building.

The present Director, Dr Mart de Groot, comes from Holland. His particular investigation is centred upon a star known as P Cygni, which is very unusual indeed. In 1600 it flared up from obscurity to the third magnitude, after which it faded; for many years now it has been dimly but perceptibly visible with the naked eye.

P Cygni is highly luminous, and is on the verge of instability. As de Groot said, 'it does not quite know whether it should hold itself together by gravity, or whether it should blow itself apart under the influence of the radiation pressure being produced inside it.' Material is being constantly expelled from the star, and there are also shells or blobs of matter which are thrown off at irregular intervals.

Spectroscopic work can tell us a great deal, but P Cygni is still something of a puzzle. Certainly it is losing mass at a tremendous rate, and it cannot last for nearly as long as a steady, well-behaved star such as the Sun; its life-expectancy cannot be more than ten thousand years at most, which is not long on the cosmical scale. What will happen to it eventually we do not know, but there is a distinct chance that it will explode as a supernova, in which case it will be very spectacular indeed.

Flare stars are also under survey by Armagh astronomers. Most of these stars are small and red; they may brighten up suddenly over a period of a minute or two, taking hours to die down to their normal state. Of course, other stars show flares, including the Sun, but these small red stars show phenomena which are much more violent. They are also spotted. On the Sun, a spot-group is never much larger in area than one-thousandth of the solar surface, but on flare stars a spot may cover up to half of the visible hemisphere. When the spotted hemisphere is turned toward us, the star is at its faintest, and we have found that it is the quickest spinners which produce the greatest numbers of flares. Evidently we are dealing with what is in essence a dynamo effect. Just as a bicycle light will brighten when the wheel is spinning rapidly (at least if it is of the old-fashioned type, as on my own cycle!), so a rapidly rotating star will be particularly prone to flares and spots.

Armagh Observatory. The Robinson dome is to the left, and the Schmidt dome to the right. The central dome holds the original tiny Armagh telescope.

A star appears only as a point of light, so that the method of investigation has to be purely spectroscopic. It has been found that the spots grow and decay over definite periods, and even re-arrange themselves, so that the light-curve will itself change with time, enabling us to draw up crude maps of the surfaces of the stars.

We are not sure where the flare stars fit into the evolutionary sequence. The obvious idea is that they are very young, and have not had time to settle down to a sober, steady existence, so that they have kept their original spin. A star is formed from a collapsing cloud of material (mainly hydrogen), and as it shrinks the rotational speed increases, just as a skater will spin more rapidly by pulling in his arms. Unfortunately things are not so straightforward as this, because the flare stars tend to move in a way which indicates old age. Proxima, the faint companion of the brilliant southern Alpha Centauri, is a flare star, and is actually the closest of all stars beyond the Sun. There can be no doubt that Proxima is as old as Alpha Centauri itself, and hence just as old as the Sun— well over 4500 million years.

Alternatively, there may be a fundamental link between rapid rotation and flare activity, in which case it is the activity which preserves the rapid spin; we believe that in the early period of our Sun's existence, mass was lost by way of a very much enhanced solar wind, and that this was responsible for slowing down the Sun's rotation to its present modest rate. But if a star is very active it will have many closed magnetic lines which will prevent matter from escaping in a stellar wind, in which case the flare stars have kept their quick spin merely because they have not been able to lose mass.

These are only some of the investigations being carried out at Armagh. Despite the lack of large telescopes at the Observatory itself, there is little doubt that the next two centuries will be as productive as those of the past.

20 ROSAT

The Rosat satellite in orbit.

In 1895 the German scientist Wilhelm Röntgen discovered X-rays. Six years later this discovery earned him a Nobel Prize. And almost a century later—on 1 June, 1990—came the launch of an artificial satellite named in his honour: Rosat, the Röntgen Satellite.

Rosat was designed to study short-wave radiations from space which can never reach ground level, because they are blocked out by layers in the Earth's atmosphere. These radiations include not only X-rays, but also ultra-violet. Rosat was designed to study the region of the electromagnetic spectrum known as EUV or Extreme Ultra-Violet, with wavelengths only just longer than those of X-rays.

There had been earlier satellites launched for short-wave studies. X-ray astronomy began in 1962 with a camera carried aloft in a rocket, and since then there have been fully fledged X-ray satellites, such as the Copernicus and Exosat vehicles. Ultra-violet has also been under scrutiny, and, as we have seen, the IUE or International Ultra-Violet Explorer is still working well. It is the EUV region which has been badly neglected, and this is where Rosat is proving so valuable.

The reason for this apparent neglect is that EUV radiations are strongly absorbed by the interstellar medium, which is made up chiefly of hydrogen. It is almost incredibly tenuous—more so than the best vacuum which we can produce in our laboratories—but it is omnipresent, and it was generally thought that it would limit our EUV range to the immediate area of the Solar System. If this were so, then there was little point in launching an EUV satellite. However, during the 1970s came a change of heart. It became apparent that the interstellar blocking was patchier and less effective than had been expected, and so Rosat went ahead. True, in some directions—notably toward the region of the sky marked by Cygnus—the blocking is a hazard, but in other directions we can see much further, and in the area of Canis Major it is possible that in EUV we can see right out of the Galaxy into interstellar space.

Rosat is moving round the Earth at a height of 360 miles, in a more or less circular path. The orbital inclination is 53 degrees, so that the satellite can pass over Britain and can be seen with the naked eye as a 'star' of about the third magnitude. It was sent up from Cape Canaveral on a Delta-2 rocket, and has an estimated active lifetime of from 5 to 7 years, though it may last for much longer (as IUE has done).

As befits its name, Rosat is essentially a German vehicle. There are two main instruments. One is the German XRT or X-Ray Telescope, while the other is the British WFC or Wide Field Camera. They point in the same direction, and study the same objects simultaneously. For obvious reasons, they always keep well away from the direction of the Sun.

The XRT is not a normal telescope. Everyone knows the penetrating power of X-rays, and the XRT makes use of the principle of 'grazing incidence', so that the incoming radiations are recorded from two sets of mirrors; the angle at which the radiations come in is a mere 4 degrees (there is some analogy here with a flat stone skipping across the surface of the sea when it is suitably thrown). The XRT is of Cassegrain design, with a 50-centimetre main collector and a focal length of half a metre. The WFC is the main link between X-rays and the more accessible regions of the ultra-violet.

First light came on 17 June, when signals were picked up at the ground station at Weilheim in South Germany. An intensive period of testing followed, and an all-sky survey began on 30 July. Before Rosat, very few EUV sources were known, but by mid-November over a hundred had been discovered, and the current total is well over a thousand.

The very first object studied was a curious star known now as WFC 1. It was already known to be a White Dwarf—that is to say a very old star which has used up all its nuclear reserves but is still very hot—though there had been some uncertainty as to whether the star, which has a visual magnitude of 13, really was the source of the EUV radiation. This point was settled by the WFC, which is able to pinpoint sources to an accuracy of one minute of arc. Only one star lay in the 'error circle', and there could be absolutely no doubt that the star and the EUV source were one and the same. The surface temperature is of the order of 35,000 degrees Centigrade, and the distance is given as 130 light-years.

Once the identity had been established, the optical spectrum was examined with the great William Herschel Telescope in the Canary Isles. At once it became clear that WFC 1 was not a single star. The spectrum was complex, and was made up of two: one indicative of the hot White Dwarf, and the other of a much dimmer Red Dwarf. This is a most unusual combination, though probably there are many similar systems waiting to be discovered.

Another White Dwarf to come under scrutiny was HZ 43, one of the half-dozen EUV sources known in pre-Rosat days. Here the surface temperature is about 200,000 degrees, making it one of the hottest stars known in spite of its small size and low total luminosity.

Also interesting is V471, in the Hyades star-cluster. Here too we have a hot White Dwarf paired with a dim red companion of spectral type K, but in this case it seems that the White Dwarf is responsible for all the EUV emission even though the red component contributes about 30 per cent of the visible light. It is an eclipsing binary system, so that when the White Dwarf is hidden by the red star the entire EUV radiation is cut off. There are other EUV sources which flicker quickly, and studies of them will certainly tell us more about stellar evolution.

A particularly exciting discovery was made in November 1990. In the southern hemisphere of the sky, unfortunately invisible from Britain, we have the Vela pulsar—the remnant of a supernova which blazed out

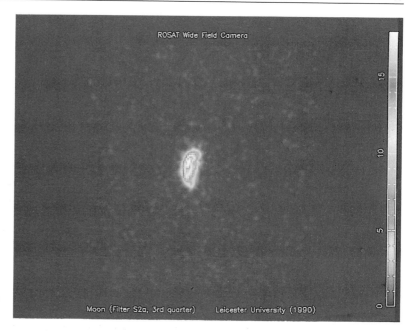

ROSAT Wide Field Camera

Moon (Filter S2a, 3rd quarter)   Leicester University (1990)

The Moon, from Rosat,
1990; the first picture of
the Moon in X-rays.

some 1500 years ago and has left a visible remnant which is also a source
of short-wave radiations. It is about 1½ million light-years away, and
measures 2° by 1° in the sky. Rosat detected it in the EUV region,
indicating a temperature of 100,000 degrees (bearing in mind that the
scientific definition of temperature is not the same as ordinary 'heat',
and the gas in the Vela remnant is very tenuous indeed). The remnant
lies more or less in the direction of the galactic plane, and the fact that
it can be detected in EUV proves that the blocking effect of the
interstellar medium really is much less than had been feared.

Even the Moon shows up in EUV. Differences in the EUV emission
over various parts of the lunar surface may well be associated with
variations in the surface material, and I am fairly sure that there are
links with the dark 'mare' areas such as Oceanus Procellarum and Mare
Nubium. As yet little has been done with EUV studies of the Moon,
but it is worth noting that this is the first time that the Moon has been
studied at very short wavelengths since the original rocket observations
of 1962. Indeed, the first X-ray camera was sent up to detect lunar
X-radiation, and though this was unsuccessful the experiment did,
fortuitously, lead on to the identification of the first discrete X-ray source,
Scorpius X-1.

Rosat, then, seems destined for a long and valuable career. It is a true
pioneer, and will fill in important gaps in our present knowledge.

21 MILLISECOND PULSARS

During the past thirty-five years there have been many *Sky at Night* programmes mentioning the Crab Nebula in some context or other, but since it comes so much into the present story I hope you will forgive a little recapitulation. The Crab lies near the third-magnitude star Zeta Tauri; it can be seen with powerful binoculars, and photographs show that it is an immensely complicated gas-cloud. Its official title is Messier 1 (the nickname of the Crab was given to it by the third Earl of Rosse, who used his great 72-inch telescope to sketch it during the 1840s). From the astronomer's point of view it is one of the most important objects in the sky, because it radiates at virtually all wavelengths from radio waves through to the ultra-short gamma-rays. It is the remnant of the supernova seen to burst forth in 1054; the distance is about 6000 light-years.

Deep inside the gas-cloud is the Crab's 'power-house', a tiny, super-dense object made up of neutrons, representing the core of the old giant star which literally blew up when it ran out of nuclear 'fuel'. This neutron star, only a few miles in diameter, is spinning round at the rate of 30 times per second. It has an intense magnetic field, and its radiation is beamed along the magnetic field lines coming out of its magnetic poles. The effect is rather like that of a lighthouse beam sweeping over an onlooker on the sea-shore. Each time the Earth passes through the Crab's beam of radiation, we receive a pulse of radio energy—hence the term 'pulsar'.

Many pulsars are now known, though only a few of them have been optically identified with faint, flashing objects. The Crab is the youngest of the known pulsars, and is spinning round at the greatest rate. It was also the first to be optically identified. In 1969 observers at the Steward Observatory in America tracked it down to a very faint, flashing object in the heart of the nebula.

A neutron star, remember, is formed by the death of a giant star. When the core collapses and shrinks, its rotation is speeded up; thus if the core had an original period of ten minutes, the resulting neutron star would spin in a thousandth of a second. It is the rotation which makes the signals repetitive. If we are in the right position, we will see the flashes and pick up the radio pulses. If not, then the beams will miss us, and we will have no way of telling that a pulsar exists.

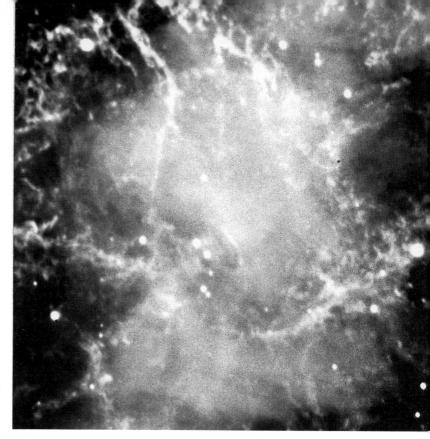

Central area of the Crab Nebula: Image with the NTT (New Technology Telescope) at La Silla. The central pulsar is the lower right one of the two brighter stars near the centre. The Wisp Nebula lies to the right of the pulsar (European Southern Observatory).

The speed of rotation cannot remain the same for ever. Just as the Earth once rotated in less than 23 hours, and now takes 24, so a newly-formed pulsar will at first spin rapidly and then slow down; in 10,000 years from now the rotation period of the Crab will have been doubled, and it will go on slowing down, losing energy as it does so, until its radiation will no longer be detectable. We can say that it will then be dead, though the process will take a very long time indeed. The Vela supernova remnant, formed as the result of an outburst 10,000 years ago (unfortunately before there were any astronomers around to record it) has already slowed to a period of 11 rotations per second, and older pulsars take even longer.

Attaching radio receivers to large radio telescopes makes it possible to record 'clicks' which represent pulsar sounds, though please note that the actual noise is produced inside the receiver, we cannot actually 'hear' noise from space! The clicks are remarkably regular, and it is not surprising that when the first pulsar was discovered, by Jocelyn Bell-Burnell from Cambridge in 1969, it was briefly thought that we might be picking up signals from a far-off alien civilization. Certainly a neutron star is a strange body. It is so dense that a pinhead of its material would weigh as much as an ocean liner.

In recent times some even more curious pulsars have been discovered, spinning at hundreds of times per second. These are the 'millisecond pulsars'. The first of them, 1937+21, was identified in 1982. It rotates 642 times per second, and is still the holder of the speed record, though it has several rivals. Translated into sound, it gives the impression of a humming-top.

In studies of this sort all the various branches of astronomy come together, and a particularly interesting case of this was described to me by Dr Phil Charles during one of my recent visits to the La Palma Observatory in the Canary Isles—site of the new William Herschel Telescope or WHT. The story began with the discovery, with the 1000-foot non-steerable radio dish at Arecibo in Puerto Rico, of 1937 + 21. It took astronomers by surprise, because it was so obviously different from a 'normal' pulsar such as the Crab remnant. It had to be a spinning neutron star, but it was also found that the radio waves disappeared for definite periods, so that the neutron star was one component of an eclipsing binary system; before long the period was fixed as 9.1 hours. The Very Large Array in New Mexico obtained an accurate position for the source, down to one-tenth of a second of arc, so that Phil Charles could begin hunting with the William Herschel Telescope.

He had no finder chart, so that he relied upon the amazingly accurate pointing of the telescope. Soon he found what he expected. Unfortunately it was within 0.6 seconds of arc from a separate star which was not associated with the pulsar, but merely happened to lie almost in the same line of sight. (This is surely a case of Spode's Law. If things *can* be awkward, they *are*.) The magnitude of the pulsar was below 24, so that very few telescopes in the world would be adequate for studies of it.

Other millisecond pulsars were soon found, and were clearly objects of an entirely new type. Apparently a millisecond pulsar is an old neutron star which has been 'spun up'. It pulls material away from its larger, much less dense red companion, and these gases are heated up, forming an accretion disk round the neutron star and increasing its rate of spin. A good way to demonstrate this is to suspend a table-tennis ball by a thread, and then blow on it, using a drinking straw. The ball will soon be spinning round quickly.

As the gases approach the neutron star they are so intensely heated that they give off X-rays, so that in searching for millisecond pulsars it is a good plan to look for low-mass X-ray binary systems. The cores of globular clusters are particularly promising. A globular cluster is a symmetrical system which may contain more than a million stars, and near the core the star population is much greater than in other parts of the Galaxy, perhaps by a factor of a million. The component stars are moving around, and there are many close encounters, so that binary systems are formed. We may therefore expect a large number of low-mass X-ray binaries.

In 1987 Professor Andrew Lyne and his colleagues at Jodrell Bank examined the globular cluster M 28 in Sagittarius, which is around 15,000 light-years away and is easy to see with a small telescope; it has a bright, glowing centre, fading off rapidly toward the edges. The initial survey was carried out in America by the VLA or Very Large Array, which consists of a number of separate radio telescopes working together

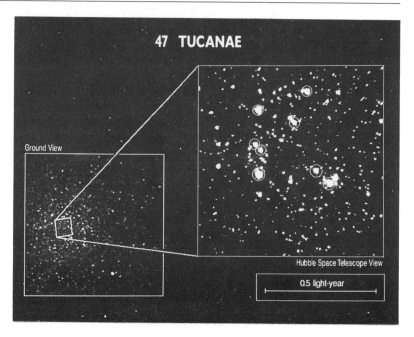

47 Tucanæ; ground and
Hubble views.

in unison. One low-mass X-ray binary was found. The VLA could detect
no pulsations, and so the Lovell Telescope at Jodrell Bank was called
in, together with the powerful Cray-XMP computer at Los Angeles.
A millisecond pulsar was duly revealed, but—surprise, surprise!—it was
not a member of a binary system. Other globular clusters proved to
be equally fruitful (notably M 4, near Antares, which is easily visible
with binoculars and is one of the closer globulars, at only 7500 light-
years). Before long nearly thirty millisecond pulsars had been found,
only about half of which had binary companions. This posed a real
problem. If a millisecond pulsar is produced by the spinning-up of an
old neutron star by material pulled from a companion, how can single
objects be explained?

The answer came in 1988 as a result of studies of the unusual pulsar
PSR 1957+20, in Sagitta, carried out by Joe Taylor and Andrew
Fruchter of Princeton University. The pulsar has a period of 1.6
milliseconds, and does have a binary companion, but this companion
proved to have a mass only 1/50 that of the Sun, which by stellar
standards is negligible. Moreover, it occulted the pulsar during each
orbit, and the observations revealed a stream of gas leaving the star in
a direction opposite to the pulsar, so that if it could be seen from close
range the general impression would be rather like that of a comet. What
seems to be happening is that the companion star is being boiled away,
or 'ablated', by the incredibly intense radiation from the pulsar, so that
in a short period of perhaps no more than 10,000 years it will have
been completely evaporated, leaving a solitary millisecond pulsar of the
same type as that found in M 28. In fact, we are dealing with a case
of cosmic murder which would fascinate even Miss Marple!

M13, the globular cluster in Hercules; Mount Wilson photograph.

Millisecond pulsars have been used to measure the mass distribution in globular clusters, and may lead to new clues about the 'missing mass' which undoubtedly exists. Moreover, some globulars are very rich in millisecond pulsars. Four have been found in the Pegasus globular M 15, which is around 50,000 light-years away, while the southern globular 47 Tucanæ contains no less than eleven.

One intriguing observation was reported on 18 January 1989 by J. Middleditch and his colleagues at the Cerro Tololo Observatory in Chile. They were searching for a pulsar in the remnant of the supernova which had flared up two years earlier in the Large Cloud of Magellan, 169,000 light-years away, and had become a conspicuous naked-eye object for a few weeks. The Cerro Tololo observers believed that they had indeed found a pulsar—but not of the type expected; the pulse rate was 1968 per second, putting it into the super-millisecond range! This was certainly curious, and all sorts of explanations were offered to account for it.

Even more curiously, the observation could not be repeated; the pulsar seemed to disappear as suddenly as it had arrived. Suspicions were aroused, and eventually it was found that the signals came not from the supernova remnant, but from part of the telescope mechanism. It is quite likely that an ordinary pulsar will eventually be detected there, but there is no sign of it as yet.

Millisecond pulsars are now being studied from many of the world's great observatories, and more are being found regularly. The fact that we can detect objects only a few miles across over distances of thousands of light-years shows how greatly our techniques have advanced during the past few decades.

22 COSMIC-RAY ASTRONOMY

Since the start of the *Sky at Night* series, in April 1957, most branches of astronomy have been covered. Plenty has been said about the very long radio waves, collected by instruments such as the 250-foot Lovell Telescope at Jodrell Bank, and about the ultra-short X-rays and gamma-rays, which have to be studied using space research methods. But for some reason or other we had never until recently discussed cosmic rays, which are not true rays at all, but high-speed particles coming from all directions all the time. March 1991 was a good time to remedy this omission, because I was joined by the new Astronomer Royal, Arnold Wolfendale, who is the first holder of the office to have made his reputation principally by studying cosmic rays.

The discovery dates back to 1912, with a balloon ascent by the Austrian scientist Victor Hess. Flying at 17,000 feet in a non-pressurized, hydrogen-filled balloon sounds (and was) decidedly hazardous, but Hess was able to show that some hitherto-unexplained phenomena were being recorded at high altitude. Originally the effects were thought to be due to genuine rays—hence the name—but before long it became clear that particles were responsible. Essentially these particles are atomic nuclei (protons, helium nuclei and so on) together with electrons and neutrinos.

The 'heavy' cosmic ray particles break up when they dash into the upper atmosphere, and also break up the atmospheric particles with which they collide, so that only the 'secondary' particles reach ground level. This is fortunate for us, since otherwise it is not likely that advanced life-forms could have developed on Earth, but it does mean that rockets and space-crafts are essential for studying cosmic radiation. There have been many purely cosmic-ray satellites by now, and many more which have carried cosmic-ray equipment.

At first there was a great deal of interest in finding out what happens when cosmic-ray particles strike atomic nuclei. Indeed, several so-called fundamental particles were discovered during this work. There is still considerable interest in these interactions, but mainly concerning those of phenomenally high energy, and in general it is fair to say that fundamental particle research is now carried out mainly with the aid of man-made accelerators, so that we are concerned with cosmic rays mainly from an astronomical point of view. Where do they come from, and what can they tell us about the universe as a whole?

Professor Arnold Wolfendale, the Astronomer Royal.

mma-ray
e at Mount
, in Arizona.

The main problem about the first of these questions is that the tangled magnetic field in the Galaxy deflects all but the very highest-energy cosmic rays. This means that their arrival directions tell us virtually nothing, because a cosmic-ray particle may wander around the Galaxy for an immense period before reaching the Earth from a direction quite unrelated to its place of origin. However, help is at hand by way of gamma-ray astronomy.

Gamma-rays really are rays, not particles; their wavelengths are even shorter than those of X-rays. They are unable to pass through the Earth's air, so once more we are forced back to space research methods, and in fact the first discrete gamma-ray source was not identified until 1969. Various special satellites have been launched, one of the latest being GRO (the Gamma-Ray Observatory), sent up by Shuttle in 1991 and subsequently named in honour of the famous physicist Arthur Holly Compton.

Gamma-rays are produced by interactions between cosmic rays and the interstellar medium, i.e. the gas between the stars. Unlike cosmic-ray particles, gamma-rays travel in straight lines, so that their arrival directions give us a reliable clue to the direction in which they originated. This means that if we know where the gas is, we can work backwards, so to speak, and find out how the cosmic-ray particles are spread around in the Galaxy. Analysis shows that the cosmic rays are particularly plentiful in the inner part of the system, so that many of the sources must lie there. (Two pulsars have also been identified as cosmic-ray sources, but there is nothing surprising about this.) Unfortunately, all these methods apply only to cosmic-ray particles at the lower edge of the energy range, and positive identifications with visible sources are still very few in number.

The numbers of cosmic-ray particles increase very rapidly at the lower end of the range of energies, and in fact about five particles pass through a man's head every second of time. For the highest-energy particles the situation is very different, and we come down to a frequency of about

The Durham University Gamma-ray Telescope, as I photographed it in 1992.

one particle per acre per decade. For detection, therefore, we have to have several instruments spread over a wide area, as found for example at the Haverah Park research centre. After all, one can hardly build a single detector with an area of a square mile or so!

At the moment it seems that most of the cosmic-ray particles which we can detect come from our own Galaxy, but we cannot be so sure about the particles with the greatest energies. The Astronomer Royal's view is that they are a mixture, with some of them coming from outside the Galaxy. There is even some evidence that some of them come from the Virgo cluster of galaxies, at least 50,000,000 light-years away; the Virgo cluster contains many hundreds of systems, and is the centre of the so-called 'super-cluster' of which the Milky Way is a member.

Yet we have to admit that we are still at an early stage in our investigations. Data collected from all over the world indicate that there are some objects in the Galaxy which are giving out cosmic-ray particles of truly enormous energy, but we do not have much idea of what exactly these sources are. Gradual progress is being made, but for the moment cosmic rays remain decidedly mysterious, even though they have been known now for eighty years.

23 BL LACERTÆ

The little constellation of Lacerta, the Lizard, is anything but conspicuous. It has no bright star; it lies in the far north of the sky, near Cepheus and Cassiopeia, and in most ways it is entirely undistinguished. But it does contain one object which is particularly fascinating: BL Lacertæ.

It is not visible without a reasonable-sized telescope, because it never becomes much brighter than the twelfth magnitude, well beyond the range of ordinary binoculars. It was first noted in 1941 by H. van Schewick, and found to be variable, so that it was classed as an ordinary irregular variable star; but as soon as its spectrum was examined, it was found to be something much more dramatic.

A normal star, such as the Sun, has a spectrum which looks like a rainbow band crossed by dark absorption lines, each of which is the trademark of one particular element or group of elements. But when astronomers examined BL Lacertæ, they had a shock. All they could see was a totally featureless band, with no lines at all. As David Allen commented, BL Lacertæ had won the first round!

Clearly it was no normal star; so could it be a galaxy, far beyond the Milky Way? Eventually a fuzzy patch was detected around it, showing that it was indeed very remote and was in the nature of a star-system rather than a star. The same was true of other similar objects which had been given variable-star designations, such as W Comæ and AP Libræ. Because BL Lacertæ was the first to be identified, it has given its name to the whole class, though most people call them simply 'BL Lacs' for short. Over a hundred are now known. Most—including BL Lacertæ itself—are strong, point-like radio sources, and quite a few are also strong X-ray emitters as well as sources of infra-red radiation.

Apparently we are seeing a jet of material coming from deep inside the nucleus of a remote galaxy. The jet is moving directly toward us at a tremendous speed, not far short of that of light. Optically the jet is extremely small, and looks like a bright speck which overpowers the light from the rest of the galaxy, rather in the manner of a quasar. With BL Lac, the prototype system, the essential studies were made in 1973 by Oke and Gunn, using the 200-inch reflector at Palomar Observatory in California, which is still one of the most powerful in the world even though it is now well over forty years old.

It was no easy matter to obtain a spectrum of the fuzz round BL Lac, but Oke and Gunn managed to do so, and were able to identify lines

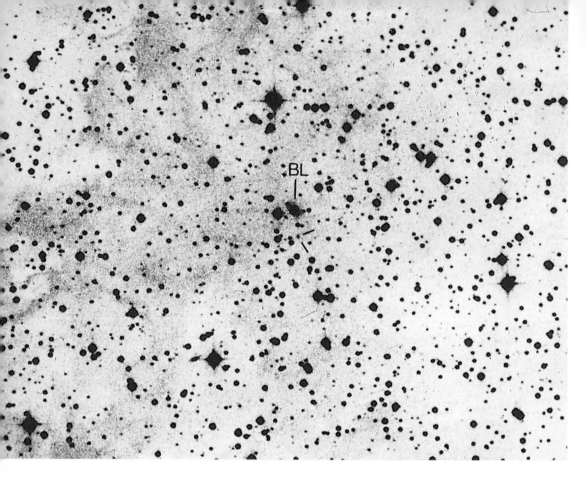

BL Lacertæ.

in it. These lines were shifted over to the long-wave or red end of the rainbow band. A red shift indicates a velocity of recession (this is the well-known Doppler effect), and BL Lac was found to be racing away at over 12,000 miles per second. This fitted in with the general picture of an expanding universe, and at a later stage new equipment was able to show BL Lac's associated galaxy quite clearly.

The jet contains highly energetic electrons, which is by no means unexpected. The overall result is emission of light by what is termed the synchrotron process, due to electrons being accelerated in a strong magnetic field. The jet theory is supported by the fact that the light coming from the spiralling electrons is obviously polarized, as is characteristic of synchrotron emission.

Much has been heard of quasars, which have now been known for more than a quarter of a century. It has started to look as though quasars, BL Lacs and very strong extragalactic radio sources are all much the same, and look different only because we see them at different angles. This idea—the so-called 'unified scheme' for active galaxies—was first proposed by Ian Browne, from Jodrell Bank, in the early 1980s, and has since been refined and extended. To show what it involves, consider a double radio source such as Cygnus A, where there are very thin, faint jets stretching from the central galaxy out into the radio lobes. If we could look 'down' one of the jets, the whole aspect would be

Lacerta, Cygnus, Lyra and Vulpecula.

changed. We would see a central spot surrounded by low-level emission, and the jets would be dominant, as they are with BL Lacertæ. Of course, there are also cases when we are not looking directly down the jet face-on at the radio source, but are viewing from an angle. The jet which is coming toward us will then look brighter than it would otherwise be, but the jet moving away from us will be weakened, and may not be visible at all. We see this with quasars, many of which have only a single visible jet although we can often detect the stationary 'lobe' emission equally on both sides of the central optical object.

Powerful radio sources are usually associated with massive elliptical galaxies, so that we would expect BL Lacs also to come from these large systems. Ian McHardy, Roberto Abraham and Carolin Crawford therefore set out to test the 'unified scheme' by taking detailed optical images of BL Lac objects to decide whether they really were associated with large elliptical galaxies. Observations of this kind are very delicate, because the BL Lac core drowns the emission from the stars in the surrounding host galaxy, but the William Herschel Telescope at La Palma proved to be equal to the challenge. It has been found that many BL Lacs, including AP Libræ, are indeed associated with ellipticals, but one, PKS 1413+135, is quite definitely associated with a disk galaxy.

Galaxies come in various forms. There are many systems of classification, but in general we still follow the basic scheme laid down by

Edwin Hubble in the 1920s. There are for example various kinds of 'ellipticals'; some are almost spherical, while others look like disks. We also have spirals, plus the strange barred spirals in which the arms seem to come from a bar through the centre of the system, while some galaxies are completely irregular.

For the moment, let us concentrate upon the ellipticals. Much depends upon their surface brightness properties, notably the way in which the brightness falls off as we move out from the centre of the system. With a BL Lac we have a bright, central, starlike core, which is the emission from the jet itself, surrounded by the faint fuzz. The procedure is to check the profile of the galaxy associated with the BL Lac object in which we are interested. With BL Lacertæ itself, the curve fits in perfectly with an elliptical system, but with PKS 1413+135 the situation is different; the surface brightness profile shows that we are dealing not with an elliptical, but with a disk galaxy.

As yet the images are not very clear, and we cannot even be sure that there are no spiral arms associated with the PKS 1413+135 galaxy, but it does look 'flat', which is rather strange. Disk galaxies are not renowned for their strength at radio wavelength, and we would not expect to find BL Lacs inside them. We must wait and see whether any more can be located.

There have been other attempts to explain BL Lacs. For example, it was once suggested that a BL Lac is nothing more than a very remote quasar which happens to lie behind some closer galaxies, so that the nearby galaxies act as lenses and make the background quasar appear much brighter than it really is, but this does not seem likely. If it were true, we would expect the BL Lacs to be distributed all round the central regions of their host galaxies, and this is not what we find. The BL Lacs are always exactly central.

It looks as though we are on the right track, but there is still a great deal which we do not yet know. As David Allen said, BL Lacertæ won the first round, and we are no more than half-way through the second.

24 THE WHITE SPOT ON SATURN

Saturn is the Ringed Planet. Of course Jupiter, Uranus and Neptune also have rings, but none of these dark, obscure systems can rival the glory of Saturn. When the Saturnian rings are wide open, as they were at the start of the 1990s, there can be little doubt that they are more beautiful than anything else in the sky.

In some ways it is a pity that the rings tend to divert attention from Saturn's globe. Remember, we are dealing with a planet which is second in size and mass only to Jupiter. The equatorial diameter is 74,194 miles; the polar diameter of 67,525 miles is less simply because the rapid rotation (less than 11 hours) makes the equatorial zone bulge out, and even a small telescope will show that Saturn is markedly flattened. The mass is 95 times that of the Earth. The overall density is less than that of water, and Saturn has a great deal of hydrogen and helium in its make-up, with only a relatively small silicate core. As with Jupiter and Neptune, though not Uranus, there is considerable internal heat, even if Saturn is not nearly massive enough to qualify as an embryo star.

On the globe we can see cloud belts, not nearly so prominent as those of Jupiter. There is nothing comparable with the Jovian Red Spot,

The 1933 White Spot on Saturn, discovered by W. T. Hay; this is Hay's own drawing of it.

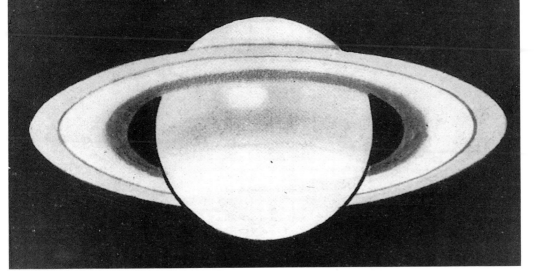

though when Voyager pictures came back from close range in 1980 and 1981 they did reveal a small, probably temporary red patch. Windspeeds are strong—indeed, stronger than those on Jupiter—so that Saturn is a very active world.

Now and then it springs surprises on us.* In 1876 Asaph Hall, probably best remembered today for his discovery of the two dwarf satellites of Mars, detected a bright white spot in Saturn's equatorial zone. It did not last for long, but another appeared in 1903, and yet another in 1933. This last spot was the one discovered by Will Hay, using his 6-inch telescope from Outer London, and it was promptly confirmed by Dr W.H. Steavenson, one of the few twentieth-century amateurs to have served as President of the Royal Astronomical Society. I heard about it on the following day, and was able to see it for myself with the 3-inch refractor which I had just acquired for the princely sum of £7.10s—and which Dr Steavenson had actually chosen for me.

Like the previous spots, this one did not last, but spread out until it had merged into the bright equatorial zone. It certainly gave the impression of having been caused by upcurrents from below the cloud deck. This was also true of the next white spot, that of 1960. Note that the spots seem to occur roughly every 30 years (we cannot expect to have any record of them before the mid-nineteenth century). Saturn takes 29½ years to complete one orbit round the Sun. This may or may not be a case of coincidence.

By 1990 a Spanish astronomer, Agustin Sanchez-Lavega, was predicting that a new spot ought to be imminent, and he was right. On 24 September two American amateurs independently detected a new outbreak. One was Stuart Wilber of Las Cruces, New Mexico, using his home-made 10-inch reflector; at once he called Clyde Tombaugh, discoverer of the planet Pluto, and Tombaugh telephoned observers at the New Mexico State University, where Scott Murrell promptly confirmed the spot with the Observatory's 24-inch telescope. The second discoverer was Alberto Montalvo of Burbank, California. He may actually have been the first to see it, but he did not report it at once, so that the main credit is officially Wilber's.

At that time I was abroad at a conference. As soon as I came home I did my best to obtain a view of Saturn, which was in the constellation of Sagittarius and therefore very low down as seen from Sussex; it skirted the trees of my southern horizon. There was also an irritating amount of cloud. Finally, on 8 October, I had my first view, using my 12½-inch reflector. And there was the white spot; I could not miss it.

The spot behaved in a fairly predictable way. By the time I first saw it the length had grown to 10,000 miles, and no doubt the brightness had decreased since Wilber and Montalvo had first seen it. By the middle of October the length had increased to 50,000 miles, and several smaller spots appeared in and near it as it faded. By the last week of October the spot had been sliced through by a thin, dark streak, with the northern

---

*See also page 65. As I am anxious to make each chapter self-contained, I hope that you will forgive a little recapitulation.

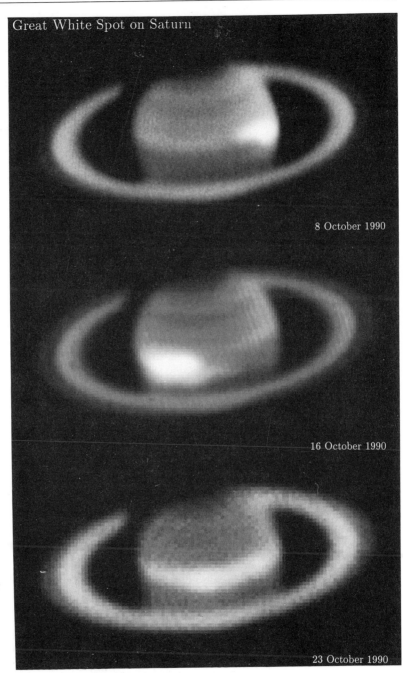

Great White Spot on Saturn

8 October 1990

16 October 1990

23 October 1990

The White Spot on Saturn: photographed by O. Hainaut and S. D'Odorico with the NTT (New Technology Telescope) at La Silla (European Southern Observatory).

part of the spot appreciably brighter than the southern. Now it had really ceased to be a proper spot, and had been transformed into a bright zone girdling the whole globe.

Ground-based telescopes had shown the sequence of events well, but were outmatched by results from the Hubble Space Telescope, which

The White Spot on Saturn, from the Hubble Space Telescope. By the time that this picture was taken, the spot had extended into a bright zone, and was not really a 'spot' any longer.

had a 94-inch mirror and was circling the Earth at a height of 300 miles in a period of 95 minutes. As everyone knew (to their cost!) the mirror was flawed, but for Solar System work it was ideal, and the pictures of the white spot taken on 9 November were staggering. After elaborate computer processing to remove the blurring due to the fault in the mirror, the results were better than anyone had dared to hope. The white spot disturbance was shown to be a twisted band of cirrus-like cloud, 200,000 miles long by 6000 miles wide, with frayed edges which formed a sort of festoon along the northern edge of the band.

It was fortunate that the Hubble Telescope was in operation, because the spot did not persist for long. By the time I was able to see Saturn again after it had come round from 'behind' the Sun, the spot had gone; the equatorial zone still looked a little brighter than usual, but that was all. We had witnessed a rare phenomenon.

What caused it? Well, it was certainly due to an uprush of material from below the clouds. Large clouds do move upward, though the lifting mechanism is rather uncertain; it could be due to the sublimation of ammonia 'blobs' which rise to a level where they are cooled sufficiently to crystallize. At least we have a good idea of the storm pattern. To quote the planetary scientist Reeta Beebe, 'When material expands northward from the equatorial zone, winds decrease rapidly, so that clouds which are trapped between these two zonal winds tend to shear, responding by forming lovely wave patterns. If we can fully interpret these patterns, we can learn more about the winds themselves.'

We cannot tell when the next white spot will appear, but observers will certainly be on the alert in 2020, after the next thirty-year interval. I doubt if I will be able to make observations myself, since I will then be 97 years old, but others will, and they could be treated to the sight of another major white spot. So despite the fascination of the rings, do not forget that Saturn's globe too merits close attention.

25 BAIKONUR

It is now many years since I paid my first visit to Cape Canaveral, the American launching base. There has never been any problem in going there, and in fact there are daily tours conducted by NASA guides. It is an impressive place, though perhaps not quite so awe-inspiring as it used to be at the time of Apollo, when the great Saturn-5 space-craft were taking men to the Moon.

The Russian equivalent of Canaveral is Baikonur in the general region of the Sea of Aral, not very far from the border with Afghanistan. Comparatively few 'foreigners' have been there. Transport is not easy; there are no roads to Leninsk, the main city, and neither are there any passenger flights. Until fairly recent times, with the coming of *glasnost*, the very existence of the rocket ground was not officially admitted. It is nowhere near the village of Baikonur; the railway station is Tyuratam, and the airfield is Krainj. You will find no details on any pre-Gorbachev map. So when the chance to visit Baikonur came to me, in the spring of 1991, I took it.

This, of course, was before the abrupt and catastrophic break-up of the Soviet Union. It was fairly clear that the situation was delicate, but that was all. I for one would have found it hard to believe that in a matter of months the USSR would be no more. The effects on Baikonur are bound to be dramatic, but I doubt whether they have yet become at all clear.

I made my way from London to the Finnish capital of Helsinki, and thence to Moscow. Then I boarded a Russian military plane for the four-hour flight to Krainj. The aircraft was not exactly luxurious (the seats were wooden, and toilets were conspicuous only by their absence), but it reached Krainj on time, and from there I drove from the airport to Baikonur, via Leninsk. Leninsk, by the way, has 100,000 people, every one of whom is associated with the space programme. To say that it is a desolate area is an understatement; all food has to be brought in by train. (I wonder whether it will retain its old name? Now that Leningrad is once more St Petersburg, I rather doubt it. Time will tell.)

The selection of Baikonur was made in the early 1950s, when the Russian space programme was just starting to accelerate. A major complex was needed. It had to be remote—otherwise, broken pieces of rocket might cause heavy damage and casualties if they came down prematurely—and it had to be as close to the equator as possible, because if a rocket is then launched in an easterly direction it can take full advantage of the Earth's axial rotation. The site also needed to have fair weather, and Baikonur has over 300 sunny days in an average year, though of course the winters are bitterly cold while the summers are fiercely hot. Moreover, the Russians wanted to keep their experiments to themselves, and a distant part of Kazakhstan seemed to be as good as anywhere else in the Soviet Union. So in 1954 or 1955 the first launch-pads were built, together with all the facilities for the workers. In less than three years Baikonur was ready to launch the first artificial satellite, Sputnik 1, which soared aloft on 4 October 1957 and ushered in the Space Age.

Baikonur is not Russia's only launching ground, but it is certainly the most important one, and most of the major projects have started from it. Nothing could be less like Cape Canaveral. Even in spring it was icy and chill; there is nothing of the bustle of the Cape, and Baikonur is much more spread-out, so that driving from one complex to another takes considerable time.

The first launch-pad is named in honour of Yuri Gagarin, the first of all space-men, who was sent up in April 1961 from this actual site (though it has been considerably modified since then). Gagarin was the archetypal cosmonaut; pleasant, good-looking and modest, as I know from the various occasions when I met him. He remains a national hero, and his statue looks down over Leninsk. Even at the time of my visit it was fairly clear that Gagarin was a great deal more popular than Lenin himself!

During my stay I was able to see an actual launch. The space-station Mir was in orbit round the Earth, and an unmanned rocket of the Progress series was scheduled to take supplies up to it. The viewing station was not unlike that at Cape Canaveral, except that it was much less crowded—despite the presence of at least a dozen bemedalled Russian generals who had come along to watch.

The Progress vehicle was a three-stage arrangement, with four strap-on boosters. The plan was that after two minutes into flight the boosters would fall away, and after nine minutes the Progress would be in orbit ready to link up with Mir. I watched with admiration, camera poised (there was absolutely no objection to my taking photographs, and everything was extremely friendly). The motors fired exactly on schedule; there was a roar, and the Progress lifted off, much more rapidly than had been the case with the last launch I had seen— Ulysses, the solar polar probe, which had been sent up from the Cape several months earlier. There was the usual glare from the rocket's exhaust, and Progress was on its way. I took several pictures of it as it gained altitude, but in a surprisingly short time it was lost to view, leaving only an empty launch-pad and a trail of drifting smoke. It was all very calm and precise, though I learned later that there had been problems with docking on arrival at Mir, and two attempts had to be made.

Mir itself was sent up from this particular launch complex. So was Gagarin himself, and also the two Vega probes to Halley's Comet, as well as the two space-craft sent to Phobos, the larger of the two dwarf

Launch of a progress rocket from Baikonur, as I saw it during my visit there. The rocket was unmanned, and was a supply vehicle for the orbiting Mir space-station.

moons of Mars. But there are other pads too, and I was able to see them, though it meant a good deal of driving because altogether Baikonur covers almost 400 square miles.

The largest complex is that of the Energia rocket, now the most powerful in the world, and the only one capable of launching Buran ('Snowstorm'), Russia's Space Shuttle. At the time of my visit there had been only two launches. The first of these, on 15 November 1988, had been successful enough, but the vibration caused by the departing rocket was so much greater than had been expected that the complex was more or less destroyed, and had to be re-built before the second Energia launch. Before going up, the rocket itself sits between two huge launching towers; in case anything goes wrong, there is an evacuation tube down which the cosmonauts can make a hasty retreat.

There was no Energia on the pad when I was there, but there was one in the assembly building, which is strictly comparable with the VAB or Vertical Assembly Building at Cape Canaveral. It is fascinating historically, because we now know that the Russians really did plan a manned flight to the Moon in the 1960s, and this was the building used at the time. There were so many failures that the whole project,

The Energia launch-pad at Baikonur.

then known as N1, was abandoned after the Apollo success, and in 1974 the building was transformed into the present Energia complex.

Energia is almost 800 feet long. At present it is used with four boosters, and can carry 100 tons into orbit; when fully completed, with eight strap-boosters, the payload could be as much as 200 tons. Only Energia can launch Buran—and I was fascinated to see Buran itself, which was being prepared for its next flight. It is essentially similar to the American Shuttle, but with differences in detail. Its huge cargo bay can carry 30 tons of payload into orbit and bring 20 tons down; the wingspan is about 80 feet, and the underside is covered with black tiles to protect the spacecraft from the intense heat generated by friction as it re-enters the atmosphere after leaving orbit. Buran's first flight was purely automatic, and judging from the film which I saw the whole mission was faultless.

Energia is vital for the Mir station; it has also sent up the various associated modules, such as the Kvants (mainly astrophysical) and Krystal. But there has been something of a surprise about Mir itself. The Russians have learned a great deal from the American experiences, and now plan to set up a sort of orbital complex rather than a single, rigidly-built space-station as envisaged by pioneers ranging from Konstantin Tsiolkovskii to Wernher von Braun.

Facing page: Energia—Russia's most powerful rocket.

American astronauts are trained in various places, but in Russia the training is done mainly at one centre, Star City, which is near Moscow. I went there after leaving Baikonur, and I was able to see a full-scale

model of Mir and its attached modules. To be candid, Star City is not particularly impressive—there is nothing of Professor Quatermass about it!—but it is extremely effective, and it is well known that the Russians, unlike the Americans, have never believed in 'show'. Until the Gorbachev era they had always done their very best to keep their space centres out of the public eye altogether.

I also heard a great deal about future Russian plans. For example, there are the Spectrum and Regatta projects. It is hoped to launch a large space telescope with a 67-inch mirror, second in size only to the Hubble. Ultra-violet space programmes were being worked out, and I was told that it was hoped to launch radio telescopes; antennæ in space can be linked with instruments on Earth according to the interferometer principle, and obtain tremendously high resolution. One radio telescope may be set up at a distance of ten Earth diameters from the Earth itself, and could in theory provide resolution ten times better than anything previously attempted. The Relict satellites will investigate the background radiation coming in from space in all directions, and neither must we forget the gamma-ray and X-ray satellites which will follow the current Russian satellite Granat into space. Even more important,

Mock-up of the Mir space-station, on display at Star City.

perhaps, will be the Lomonosov astrometric project (named in honour of Russia's first great astronomer), which aims to include 400,000 stars down to the tenth magnitude and to be more extensive and more accurate than the Hipparcos project now under way.

So far as the planets are concerned, it seems that the main concentration will be upon Mars rather than Venus. In March 1994 it is hoped to send a probe to Mars and put a balloon into the thin Martian atmosphere, so that it will drift around carrying a suspended instrument package— giving a strange impression of a snake; it will descend at night, when the temperature falls, and rise again on the following morning, taking its package to a different location. There is also a great deal of interest in Mars Rovers, which are 'crawlers' and are really developments of the Lunokhods which so successfully moved around the surface of the Moon. A Mars Rover should be able to cross ravines and clamber over rocks; one design which I saw on film gave a remarkably realistic impression of a crawling centipede. Next, no doubt, will come a sample and return mission—and finally the first manned flight to the Red Planet, which must surely be an international one.

Certainly the Russians have made tremendous progress since 1957, even though there have been inevitable setbacks (notably N1), and they now regard astronomy and space research as simply two branches of the same science. 'We do not make any divisions for our investigations,' commented Professor Alexander Boyarchuk, the then President-elect of the International Astronomical Union. 'To study a galaxy, for example, means taking data from space, from radio telescopes, from optical instruments, infra-red telescopes and many others. All these data are combined to build up the model of the galaxy. Some countries still separate their scientists into astronomers, solar physicists, space researchers and so on, but in the Soviet Union we put them all together— one science, in no way divided.'

But . . . today there is no Soviet Union; it has passed into history, and the entire situation has changed beyond all recognition. Baikonur, for example, is in Kazakhstan, now an independent state, while Star City is in Russia proper. So what will happen next? I can only say that I wish I knew. There have been suggestions that Mir will be abandoned, and Buran shelved; unquestionably there are pressing financial problems, and it is no longer possible for some official in the Kremlin to write an open cheque. As yet (July 1992) it is too early to make any reliable forecasts. I very much hope that the programmes will continue, and that there will be greater collaboration between East and West than has ever been possible before, and there seems no logical reason why this should not be so. Of course space research is not cheap, but the benefits to mankind are enormous, and it is worth noting that a probe to Mars costs rather less than a couple of nuclear submarines.

26 A GALAXY IN CREATION?

Records are made to be broken. The latest to go concerns the most luminous object ever observed. It now seems that the title belongs to a galaxy, known only as IRAS F 10214 + 4724, which has 300 million million times the power of the Sun!

The story really began in 1983 with the launch of IRAS, the Infra-Red Astronomical Satellite, which was one of the most successful of all space-craft even though it operated for less than a year. It mapped the entire sky at infra-red wavelengths, and in particular it recorded many galaxies. These 'IRAS galaxies', as they were called, were often immensely distant, and could therefore be used to catalogue the more remote parts of the universe. One of the teams concentrating on this research was based at Queen Mary and Westfield College in London, and was led by Michael Rowan-Robinson.

In 1989 the California Institute of Technology released a new, deeper catalogue of infra-red sources, including about 80,000 galaxies spread over the sky. Rowan-Robinson arranged for a consortium of astronomers at Queen Mary, Cambridge, Oxford, Durham and CalTech to apply for time on the Isaac Newton and William Herschel Telescopes on La Palma, to study a sample of 1400 of these infra-red galaxies. The selected areas of the sky were chosen so as to make the best possible use for coverage by IRAS and minimum absorption by interstellar dust. Altogether it took forty nights of telescope time—and the William Herschel, with its 165-inch mirror, is one of the most powerful telescopes in the world today.

During a visit to La Palma on behalf of *The Sky at Night* I was able to watch Rowan-Robinson and his colleague, Tom Broadhurst, at work. Though there was no means of knowing at the time, this was just before the great discovery. In the control room of the William Herschel Telescope, Rowan-Robinson checked the finder chart; Broadhurst typed in the co-ordinates of the source, and when the telescope was correctly positioned the field was examined with the television guider. (How unlike the cold, laborious work of men such as Edwin Hubble, less than half a century ago!) The candidate object was identified optically, and the telescope moved slightly to bring it on to the slit of the spectrograph. Integration began immediately, and when it was completed the data appeared automatically on the two-dimensional display. They were then

IRAS: The Infra-Red Astronomical Satellite, in orbit.

transferred to the data reduction computer, and the spectrum was analysed and displayed on the television screen.

A few nights later, just after I had left, it was the turn of the source IRAS F 10214+4724. There were several candidate objects visible on the Palomar Sky Survey Chart, two of which were provisionally lettered A and C. Which was the infra-red source? On the night when I had been there Object A had been studied, but proved to be an ordinary star. When Object C was examined, three nights later, it was found that there was another faint spectrum visible, with an unusual pattern of emission lines. Clearly it had a high red shift, and so had to be very remote, but for some time efforts to track it down were unsuccessful. When an image of the field was obtained with the Palomar 200-inch reflector, at Rowan-Robinson's request, it was found that there was a new, faint object between Stars A and C. It was provisionally labelled F.

At about the same time, a map of the same field at radio wavelengths was obtained with the VLA (Very Large Array) in New Mexico. This showed that the only radio source in the field was within one arc-second of Object F. The astronomers were now certain that Object F really was the infra-red source—the error box was a mere 30 arc-seconds—but the red shift was extraordinarily large; it amounted to 2.286, so

Dome of UKIRT (the United Kingdom Infra-Red Telescope) on Mauna Kea.

that the spectral lines were being shifted toward the red by over 200 per cent.⋆ The derived distance was of the order of 16,000 million light-years, and the total infra-red power worked out at 300 million million Suns, or 30,000 times the output of our Milky Way galaxy. Yet only 1 per cent of this power was radiating at optical and ultra-violet wavelengths. The remaining 99 per cent was all in the infra-red.

The immediate interpretation was that the power source is shrouded in dust, which absorbs the optical and ultra-violet radiation being sent out and re-emits it in infra-red. In other words, the infra-red radiation recorded by IRAS comes from a huge dust-cloud perhaps 100,000 light-years across. The object gave every impression of being a galaxy, because the optical image looked extended and the total size was about right, but it was clearly something very much out of the ordinary.

The next step was to call in UKIRT, the United Kingdom Infra-Red Telescope on the summit of Mauna Kea in Hawaii. Object F was imaged at near infra-red wavelengths, and these images showed that the object really was extended, though attempts to obtain results at millimetre wavelengths (using the James Clerk Maxwell Telescope or JCMT on Mauna Kea) and at 10 micrometres (using the MMT or

---

⋆The 2.286 is a measure of the term usually denoted by z. I hope you will forgive a little mild mathematics! Let λ represent the observed wavelength of the spectral line, and λ_0 the wavelength which it would have if it were not in motion (i.e. λ_0 is the 'laboratory wavelength'). Then

$$z = \frac{\lambda - \lambda_0}{\lambda_0}.$$

Quite straightforward, you will agree!

Multiple-Mirror Telescope on Mount Hopkins in Arizona) failed; in view of the extreme faintness of the target object, this was hardly surprising.

There are two plausible explanations for the nature of the power source:

1. The source is a very luminous quasar in the centre of a dusty galaxy, hidden from our direct view. This would be most interesting, because it would mean that the quasar had 'switched on' comparatively recently, probably less than a million years before the time that we are now seeing the galaxy. It could not be older, as otherwise it would have had enough time to blow the dust and gas out of the centre of the galaxy. The main importance of this is that we have never before identified a quasar so early in its career.

2. The energy comes from an enormous burst of massive star formation within the clouds of dust; about a thousand million stars would be needed. In this case, the energy requirements would be so great that we would probably be seeing the process of formation of a very massive galaxy.

Group of four galaxies in Leo: NGC 3185, 3187, 3190 and 3193. The first three are spirals, and the last an elliptical system. Photograph with the 200-in Hale reflector at Palomar.

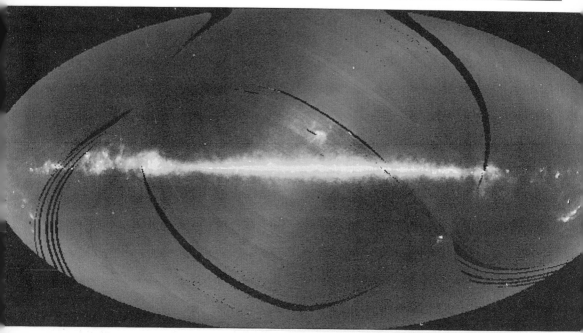

The Infra-Red Sky: Image assembled from IRAS pictures in 1983. The bright horizontal band is the plane of the Milky Way; the centre of the Galaxy is at the centre of the picture. The hazy S-shaped feature crossing the image is the faint heat emitted by dust in the Solar System. Celestial objects shown include regions of star formation in Ophiuchus (above the galactic centre) and Orion (the two brightest spots below the plane at far right), The Large Cloud of Magellan is the relatively isolated spot below the Milky Way, right of centre. Black stripes are regions of the sky not scanned by IRAS.

According to accepted theory, we are seeing the galaxy as it used to be about 2000 million years after the Big Bang, when the entire universe came into existence. Since the remotest known quasars date back to around one thousand million years after the Big Bang, and since quasars are believed to be contained inside galaxies, this means that at least some galaxies must have been formed by that time. There are sound arguments for believing that the disks of spiral galaxies were produced considerably later than elliptical systems and the central bulges and the haloes of spirals, so it could well be that we are watching the formation of the disk of an enormous spiral similar to our own, apart from being much larger.

Searches for other comparable galaxies are now going on, but they are very difficult, because Object F was one of the very faintest sources detected by IRAS, and at optical wavelengths it was not visible at all on the Palomar Sky Survey.

We do need to prove that the infra-red radiation does in fact come from the dust. This should be possible from radio and sub-millimetre observations. Secondly, we must decide whether or not there is a quasar hiding coyly inside the dust; near infra-red observations give the best chances here, because they are better able to penetrate the dust-cloud. Thirdly, we need to find some more examples of this strange phenomenon, so that we can work out whether or not it represents a critical stage in galaxy evolution.

Much remains to be done. Yet it is possible that IRAS F 10214+4724 may turn out to be the most significant extragalactic object to be found since the discovery of the first quasar, 3C-273, in 1963.

27 THE POLE STAR

Probably the most famous star in the night sky is Polaris, leader of the constellation of Ursa Minor, the Little Bear. It is not particularly brilliant, and in fact it comes only 48th equal in the list of the brightest stars; it owes its pre-eminence to the fact that it lies within one degree of the north pole of the sky.

The obvious result of this is that Polaris seems to remain almost stationary, with the entire sky turning round it once in 24 hours. Moreover, its apparent altitude above the horizon is also the observer's latitude. From the north pole the altitude will be 90 degrees, and the latitude of the observer is 90 degrees north; from London the angle is approximately 51 degrees, and so on. From the equator the altitude is 0 degrees, so that Polaris is on the horizon (or would be so, if it were precisely at the polar point). From southern latitudes it can never be seen at all.

Polaris has been of immense value to navigators throughout the ages. Measure its altitude, apply a small correction to allow for the fact that the star is slightly away from the true pole, and you know your latitude on the surface of the Earth. Finding your longitude is a much more difficult problem, but that is another story altogether!

The declination of Polaris for epoch 2000 is +89° 15' 51". The actual pole lies close to the star in the direction of Alkaid, in Ursa Major. But Polaris has not always been the pole star, because of the effect known as precession. As the Earth spins, it 'wobbles' slightly in the manner of a gyroscope which is running down, and this changes the direction of the axis in space; over a period of 25,800 years the pole describes a circle some 47 degrees in diameter. At the time when the Egyptian Pyramids were built the pole star was Thuban or Alpha Draconis, in the constellation of the Dragon. Polaris then took over the position of honour, and will be at its closest to the pole (27' 31" away) in the year 2012, after which it will start to draw away once more. Its place will be taken by the third-magnitude star Alrai or Gamma Cephei, and then, around AD 12,000, by the brilliant blue Vega. Eventually, after the full cycle of 25,800 years, Polaris will again be the pole star.

A time-exposure taken with an ordinary camera will show that Polaris is not exactly at the pole. The stars will leave trails on the photographic plate as they are carried round by virtue of the Earth's rotation; Polaris itself will produce a very short, curved trail. If it were truly polar, it would appear as a dot.

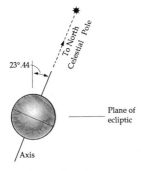

Inclination of the Earth's axis of rotation.

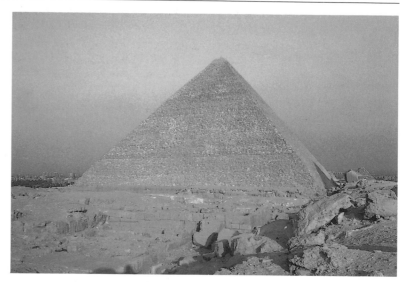

The Great Pyramid, as I photographed it in 1988.

Once Polaris has been identified, it is easy to locate again, because it does not move appreciably; it can be found by using the two Pointers, Merak and Dubhe, in Ursa Major (the Great Bear, containing the famous pattern known as the Plough, King Charles' Wain, or—in America—the Big Dipper). A line from Merak through Dubhe, and prolonged, will show the way to Polaris. It is then easy to identify the rest of Ursa Minor, which looks like a rather dim and distorted version of the Plough curving down over the Great Bear's tail. The only other fairly bright star in the constellation is Kocab, or Beta Ursæ Minoris; it is of the second magnitude, and is decidedly orange in colour. Use binoculars to look at Polaris and Kocab in succession, and you will see the difference at once.

Inevitably there is a legend about the Bears. It is said that Ursa Major was originally a beautiful princess, whose loveliness exceeded that of Juno, queen of Olympus. Juno was not noted for her kindly nature, and in a fit of jealousy she turned the princess into a bear. Years later the princess's son, Arcas, was out hunting when he encountered the bear, and not realizing that it was his mother (after all, why should he?) was about to shoot it when Jupiter intervened. He turned Arcas into a bear also, and snatched both animals up by their tails, lifting them into the sky. That is why both the Bears have tails of decidedly un-ursine length!

Now let us consider Polaris itself. It is of spectral type F8, which means that its surface is slightly hotter than that of the Sun. In theory it is slightly yellowish, though to me—and I am sure to all other visual observers—it looks white. The luminosity is 6000 times that of the Sun, and the distance is approximately 680 light-years, so that we are now seeing it as it used to be in the reign of Edward II—that is to say, about the time of the Battle of Bannockburn.

The mean magnitude is 1.99, but Polaris is very slightly variable, and is classed as a star of the Type II Cepheid or W Virginis type. The changes are not marked enough to be noticed with the naked eye, and according to Canadian astronomers the pulsations are not as pronounced as they were a few years ago; it seems that they started to die down in the early 1970s. Whether the pulsations will stop altogether in the foreseeable future is not known. If they do, we will have been lucky enough to catch a star undergoing a definite change in its evolution.

At the moment Polaris is approaching us at the rate of about 10½ miles per second, but I assure you that there is no fear of a collision; both Polaris and the Sun are moving round the centre of the Galaxy, and there will never be anything like a 'close encounter'.

Polaris is not alone. A small telescope will show that it has a faint companion, of magnitude 9, at a separation of 18.4 seconds of arc and a position angle of 218 degrees. It has been claimed that the companion can be glimpsed with a 2-inch refractor, though I personally find it none too easy even with a 3-inch (keener-eyed observers will certainly do better). It has been known for a long time, and was recorded by William Herschel as long ago as 1780. It and the bright star are definitely associated, but their orbital period round their common centre of gravity must amount to many thousands of years; the true separation is at least 200 times the distance between the Earth and the Sun, i.e. around 200,000,000,000 miles. Like Polaris itself, the companion has an F-type spectrum, so that it is hotter than the Sun; faint though it looks, it is considerably more powerful than the Sun, probably by a factor of about 6.

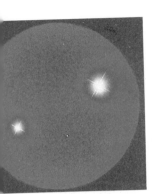

Mizar and Alcor: A representation by Paul Doherty. Ludwig's Star is shown.

Slight changes in the radial velocity of Polaris—that is to say, its motion in the line of sight—have led to suggestions that there may be a third star in the system, but it has never been seen, and must be at least five magnitudes fainter than the main star. Calculations made by Elizabeth Roemer in 1955 indicated that the unseen star is about 300,000,000 miles from Polaris, which is too close for it to be easily detectable from Earth. I must stress, however, that we have no definite proof that it exists at all.

Southern navigators always bemoan the fact that there is no comparable south polar star; the nearest naked-eye claimant, Sigma Octantis, is only of magnitude 5.5, so that the slightest mist or haze will conceal it. Moreover, it is much less powerful than Polaris, with luminosity no more than 7 times that of the Sun, and it is almost a degree away from the true south pole of the sky. So we in the northern part of the world have every reason to be proud of Polaris; it is always there for navigators to use, and it will remain the north polar star for many centuries yet.

28 REPORT ON HUBBLE

In the early part of August 1991 I visited the Space Telescope Science Institute at Baltimore, in Maryland. I had been there before, at the General Assembly of the International Astronomical Union in 1988. Then, we were eagerly awaiting the launch of the HST, or Hubble Space Telescope. Now, it was in orbit and sending back magnificent results, but it was not without its critics.

The HST has a 94-inch or 2.4-metre mirror, so that it is virtually the same size as the Hooker reflector at Mount Wilson, which between 1917 and 1948 was not only the largest telescope in the world but was in a class of its own. The HST was launched on 24 April 1990 into a 360-mile orbit which takes it right round the Earth every 95 minutes. In space, above the obstructing atmosphere, seeing conditions are perfect all the time, and it was confidently predicted that the HST would out-perform any Earth-based telescope. In many ways it actually does so, but there is a flaw: the mirror was wrongly made. During the final figuring, a vital piece of testing equipment known as the null corrector was wrongly assembled. The result is that the mirror has a curvature which is too shallow, with a total centre-to-edge error of around two micrometres, which is about one-fiftieth the thickness of a human hair. This produces an effect known as spherical aberration; light-rays hitting the mirror edges are brought to focus at a point over an inch away from the focal point of light-rays hitting the mirror centre.

How could this have happened? Frankly, it has to be put down to human error. Three tests were made. Two of them showed the fault—but the fault itself was so gross that the tests were rejected as being unreliable . . . Of course it should never have occurred, and the spherical aberration problem means that the telescope can never do all that had been hoped. Moreover, the solar panels which power the telescope show an unpleasant tendency to flap. When they pass from sunlight into shadow, or vice versa, the temperature changes make them vibrate, and this affects the delicate instruments. Finally, the gyroscopes which stabilize the telescope are proving to be erratic, so that all in all Hubble has plenty of problems.

This has led some people to call that 'The telescope has failed!' and to write it off as the most expensive fiasco in scientific history. Yet in fact nothing could be further from the truth. Obviously it is disappointing; it is not possible to re-load the telescope into the Shuttle and bring it home for repair, as originally had been hoped, and carrying out modifications in space will be a risky business. The best that can

The Hubble Space Telescope, pre-launch, as I photographed it at Cape Canaveral.

Release of the Hubble Space Telescope, from the bay of the Shuttle.

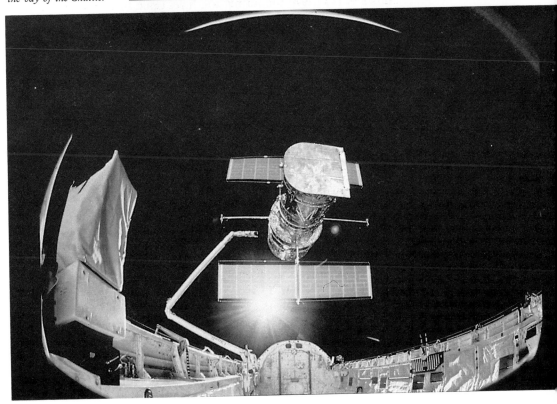

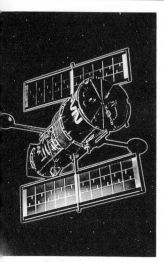

The Hubble Space Telescope.

be done will be to fit correcting lenses on to the cameras, so that they can compensate for the spherical aberration of the main mirror. Some programmes will have to be abandoned, notably the search for planets of other stars, but plenty of work remains.

For example, the Wide Field and Planetary Camera has been sending back pictures of our neighbour worlds which almost rival those sent back by space-craft. Mars shows up in amazing detail. On 13 December 1990, when the planet was 53,000,000 miles away, pictures were taken through red, green and blue filters, and the results are better than anything obtainable from ground level. The Martian north pole was obscured by a thick canopy of icy clouds; the dark basaltic rock of the V-shaped Syrtis Major showed up beautifully, as did the ochre Arabia Planitia, the basin of Isidis, and the even deeper basin of Hellas, which—for once!—was not filled with cloud. Jupiter and Saturn were also imaged, and indeed the great white spot on Saturn might well have been discovered by the HST—but the first picture was taken when the spot was on the far side of the planet, and out of view. Subsequently a magnificent series of pictures was obtained, carrying the sequence of events through to the point where the spot had been extended into a bright region girdling the globe and had lost its separate identity.

Pluto was next on the list, and for the first time it was shown as clearly separate from its companion, Charon. An obliging comet, Levy's, was shown in detail; on 27 September 1990, when the comet was 100,000,000 miles away, it was found that the light came mainly from solar rays reflected from the cometary dust expanding away from the icy nucleus.

But, of course, HST's main rôle was in studying objects far beyond the Solar System. Some of the targets lie in our own Galaxy. The Orion Nebula was superbly shown, with its 'window-curtain' structure; even from a range of 1500 light-years, the thin sheet of gas at the edge of the nebula gives the impression of a window curtain separating the diffuse high-temperature interior from the cooler adjacent dust-cloud. Sulphur emission comes from the region where the starlight is, so to speak, boiling off from the face of the denser cloud in which fresh stars are being born.

Closer to us is the symbiotic variable R Aquarii, which is a binary system of unusual type. There are filamentary features at least 250,000,000 miles long, representing geyser-like structures twisted by the force of the outburst from the star and channelled upward and outward by powerful magnetic fields. Also of special interest is the erratic variable Eta Carinæ, which once outshone every star in the sky apart from Sirius, and is associated with a strange, unique nebula which has been beautifully imaged by the HST.

On the edge of our main Galaxy, some 15,000 light-years away, lies the globular cluster 47 Tucanæ. Like many globulars it is a symmetrical system, containing at least a million stars which are so closely packed toward the centre that they are difficult to resolve. The HST has been equal to the task, and for the first time we can really see into the heart of 47 Tucanæ. It had been expected that there would be indications of a central black hole, but none has been found either in 47 Tucanæ

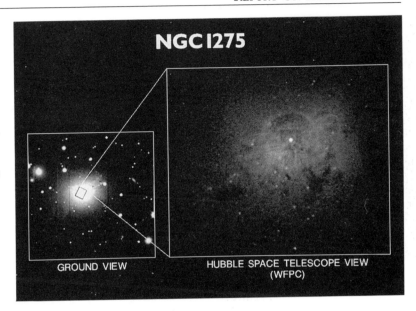

Globular clusters at the core of NGC 1275 (comparison with ground-based view): (left) A ground-based image of the giant elliptical galaxy NGC 1275, taken with the 4-metre telescope at Kitt Peak National Observatory. This peculiarly shaped galaxy lies at the centre of the Perseus cluster of galaxies in the northern hemisphere. (right) A NASA Hubble Space Telescope Wide Field/ Planetary Camera image of the central portion of NGC 1275. The HST's high resolution reveals individual clusters of stars that appear as bright blue dots. These globular clusters contain young stars—which is unusual because most globular clusters are made up of old stars. Photo: J. Holtzman/ NASA.

or in other globulars, so that some of our theories may have to be revised. However, luminous blue stars nick-named 'blue stragglers' have been seen, and over twenty have been counted in 47 Tucanæ alone. Probably they are not young, as most blue stars are, but have been produced either by interactions between two members of a binary system or by actual merging of stars which have collided.

What about systems well beyond the Milky Way? One of the nearest, at a distance of 169,000 light-years, is the Large Cloud of Magellan, easily visible with the naked eye and looking rather like a detached portion of the Milky Way itself (though, sadly for European astronomers, it lies not far from the south celestial pole—as also does the Small Cloud, with 47 Tucanæ almost silhouetted in front of it). The HST has taken magnificent pictures of the ring of material surrounding the supernova which flared up in the Large Cloud in 1987. This material was not ejected during the outburst, but was an earlier ring shed by the star when it threw off its outer layers. Also in the Large Cloud is the young star-cluster R136, which lies in the great Tarantula Nebula 30 Doradûs. R136 had been studied by ground-based telescopes, but only the HST can resolve it, showing the individual stars and giving us new insight into the earliest stages of a star's life.

Remote galaxies, of course, are prime targets for the HST. One of these is NGC 1068, which looks visually like a normal barred spiral. It is around 30,000,000 light-years away, and is the closest active galaxy of its type, with a core shining as powerfully as at least a thousand million Suns. Because the core brightness varies quickly, we can tell that the energy-producing region can be no more than a few light-days across, and this in turn indicates the presence of a super-massive black hole. Hubble's camera has resolved the inner part of the core, showing in detail how the black hole affects its surroundings through jets, stellar

Gravitational lens effect: the 'Clover leaf' Galaxy 2237+0305. Hubble Space Telescope (Faint Object Camera), 1990.

winds and ionizing radiation. An intense beam of radiation is being sent out by material which is about to be sucked into the black hole; the jet lies in the main plane, and ploughs through dust and gas to carve out a cavity of ionized gas, so that it acts rather in the manner of a blow-torch. The Faint Object Spectrograph also suggests that there is dense, high-velocity plasma near the black hole, but we do not yet know whether it is material ejected from near the hole or simply gas-clouds which have been caught up in the violent stellar wind and are flowing outward. However, the clouds—some of which are no more than ten light-years across—do show signs of being compressed by a jet of plasma from the core.

Jets from galaxies are not uncommon, but as yet we do not pretend to understand them at all well. The most celebrated of all the jets is associated with the giant galaxy Messier 87, the main member of the Virgo cluster. The Hubble results show that the jet is convuluted, rather like a corkscrew. Then there is the galaxy 3C-66B, which when examined in ultra-violet is seen to have a 10,000 light-year long plasma jet, with a distinctive, twisted, ladder-like structure. Again we have to assume that the jet is due to some violent object in the galaxy's core, and surely this can only be a black hole. The same is true of the immense jet, 30,000 light-years long, sent out from the core of the radio galaxy PKS 0522136; this is also one of the few jets to show up in visible light.

What about objects in the depths of the universe? Here the spherical aberration is certainly a serious handicap, but a great deal can still be done. I have already mentioned the 'Einstein Cross'; the picture shows a central blob flanked by four others, and by now we have been able to interpret it. The central blob is a galaxy, while the other four are images of the same quasar, which lies directly behind the galaxy and whose light is therefore split up gravitationally into four images. This is a classic example of the lens effect, predicted by Albert Einstein long before the idea of launching a space telescope had ever been considered.

These are only a few of the results to date. Despite the fault, the telescope is a technological triumph, and so are its supporting programmes. For example, there is the Guide Star Catalogue, which contains accurate brightness and positional measurements for 19,000,000 objects, of which 15,000,000 are used for aiming and controlling the telescope. The catalogue was assembled from 14,777 photographic sky plates.

Obviously the results from the telescope have to be sent for analysis, and this is not so straightforward as might be thought; remember, the HST takes only 95 minutes to go right round the Earth. The light signals are converted into digital signals, and these are routed through the Tracking and Data Relay Satellite (TDRS) which is in a geostationary orbit round the Earth at a height of 24,000 miles. The TDRS ground station at White Sands in New Mexico then relays the data, via a domestic communications satellite, to controllers at the Goddard Space Flight Center at Greenbelt in Maryland. Eventually the results arrive at the Space Telescope Science Institute in Baltimore.

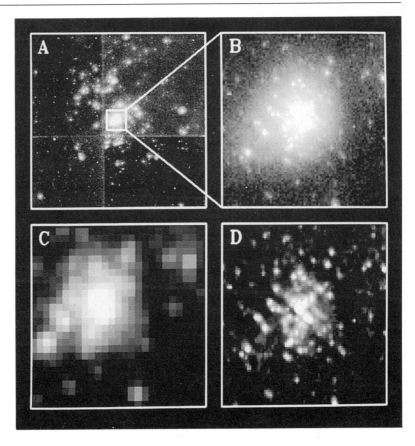

Four views of the star cluster in the Tarantula Nebula (30 Doradûs) in the Large Cloud of Magellan: Panel A (upper left) is a portion of a photograph made with the Wide Field/ Planetary Camera on HST on August 3, 1990. The camera photographed four adjoining sky regions simultaneously, which are here assembled in a mosaic. Panel B (upper right) is an enlargement of the central portion of the HST photograph, which was made in violet light. It shows the compact star cluster R136, which consists of very hot and massive young stars. The star images have bright cores that are only 0.1 arc seconds wide, allowing many more stars to be distinguished than in previous ground-based telescopic photos. Panel C (lower left) is a photograph of the same region as Panel B, both obtained with the Max Planck 2.2-metre telescope at the European Southern Observatory in Chile by Dr Georges Meylan. The star images are 0.6 arc seconds wide. Panel D (lower right) shows how computer processing of the HST image in Panel B has sharpened its appearance. The undesirable fuzzy halos around the stars as seen in Panel B are substantially reduced. Photo: HST and Ground Based Telescope.

Another great advantage of the HST is that it can operate at ultra-violet wavelengths, and its equipment is more sophisticated than that in the remarkably long-lived IUE or International Ultra-violet Explorer. For example, Douglas Duncan has been looking for traces of boron in very old stars which lie in the halo of the Galaxy, and are believed to date back to a very early period in the story of the universe. On Hubble, the High Resolution Spectrograph has to be used; the lines of boron are in the deep ultra-violet, and are therefore not accessible from ground level, while they were too delicate for the instruments carried in IUE. Boron can be produced by cosmic rays, but also by the remnant of the Big Bang which produced the universe which we know today. Duncan has found signs of 'big bang' boron, and this could give vital clues about the very beginning of the story.

Then, too, there is the problem of planets orbiting other stars. Hubble's spherical aberration means that the original plan—hiding the target star by an occulting disk, and searching for faint specks of light nearby—has had to be abandoned, but there are other ways of tackling the problem.

One star of special interest is Beta Pictoris, in the little southern constellation of the Painter, which is 78 light-years away, and is considerably hotter and more luminous than the Sun. In 1983 it was examined with IRAS, the Infra-Red Astronomical Satellite, and found to be associated with cool, possibly planet-forming material. Subsequently this material was actually photographed by astronomers at the Las Campanas Observatory in Chile, using the 100-inch Irénée du Pont reflector. A gas-cloud had also been detected from the IUE or International Ultra-violet Explorer satellite, and so the Hubble team decided to investigate further.

Their aim was to use the High Resolution Spectrograph to record the movements and distribution of certain types of iron atoms—those which have lost one electron, and are therefore said to be singly ionized. This information would lead on to a knowledge of the composition and distribution of other materials in the gas-cloud. Two observations were made, separated by 23 days, and definite changes were found.

It now seems that there really is a large disk, with a diameter of about 100 astronomical units, made up of solid particles; the gas disk is much smaller, extending out to only a few astronomical units from the star. The gas is made up of a stable diffuse disk, an inner disk which is slowly spiralling inward, and isolated gas-clumps which are also spiralling inward toward Beta Pictoris at speeds of up to 120 miles per second. As the gas-clumps pass in front of the star, the ultra-violet spectrum shows substantial changes, and it was these which were detected by the HST on 12 January and 4 February 1991. All three gaseous components seem to be embedded in a more rarefied gas-cloud which is diffusing outward, possibly because of the radiation pressure from the star. The origin of this overall cloud is unclear. It may be due to the slow breaking-up of solid particles in the disk, or it may have evaporated from cool, perhaps comet-like objects which have already formed.

In every way Beta Pictoris is a fascinating star. It seems likely that it is the centre of a protoplanetary system—or have planets already formed there? If so, are any of them like the Earth? These are questions which the HST cannot answer, but at least it can guide us along the right lines of investigation.

Edwin Hubble died in 1953. Nobody could then have foreseen that within forty years after his death a great telescope named after him would be circling the world, and sending back data far beyond the capability of any ground-based instrument. When history comes to be written, the HST will be remembered not because of the fault in its mirror, but as the telescope which opened up a whole new era of research.

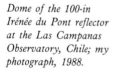

Dome of the 100-in Irénée du Pont reflector at the Las Campanas Observatory, Chile; my photograph, 1988.

29 BLUE SKIES?

Over the past few years, many people have come up to me and asked, plaintively: 'What has happened to the stars? I can't see them nearly as well as I did when I was young; why have they faded?' The answer, of course, is quite straightforward. There is no difference in the stars, but we see them much less clearly because the skies are no longer as dark as they used to be. Light pollution has become a real menace.

It is not confined to Britain—far from it! In California the 100-inch telescope at Mount Wilson, for so long the most powerful in the world, has been mothballed (only temporarily, we hope) because of the increased glow from the lights of Los Angeles. The whole situation is becoming worse all the time, and will continue to deteriorate unless we do something about it soon.

The first, and most important, point is that astronomers do not want to 'put out the lights' and make the streets and countryside darker. This seems to be the general view among members of the public, but it is quite wrong. We have to accept that law and order in Britain have broken down, more or less completely, so that dark streets are dangerous. To quote David Crawford, President of the International Dark Sky Association:

'What we want is for people to use the light, and not waste it. If we put half the light that we're generating up into the sky, it does nothing for safety or security. It simply wastes money.'

It is tempting to suggest that bright lights are useful in curbing crime, but this is so only if the lighting is suitable. Glare is the main enemy. Everyone must be familiar with the globe lights which dazzle the oncomer and cast a pool of darkness directly below in which any would-be attacker can lurk. Illuminate a building with permanent floodlights, and the burglar knows exactly where he can go. On the other hand, a security light which flashes on only when anyone walks into its beam really is a deterrent.

The 100-inch Hooker reflector at Mount Wilson.

Fortunately, officials are becoming aware of the situation. In Tucson, Arizona—one of the main astronomical centres of the world—new street lights have been installed which cut out the glare, so that from the centre of the city you can now see the Milky Way. In Britain, the Institute of Lighting Engineers is very helpful. Even more importantly, the problem has now been taken up personally by the Minister of the Environment, Mr David Maclean. In the summer of 1992 he invited me to go and see him; we had a most useful discussion, and he then gave his endorsement to a series of recommendations which had been drawn up by the Dark-Sky Committee of the British Astronomical Association and approved by the Institute of Lighting Engineers. Broadly speaking, the recommendations are:

1. Switch off all brilliant lights which are not really needed, such as advertisements, after midnight.
2. Direct all lights downward, not up. With existing lights which cannot be altered, use shields and baffles to cast the light on to the places where it is wanted.
3. Avoid glare at all costs, which means dispensing with globe lights and other atrocities.

These recommendations really work. Nobody is suggesting that all existing bad lighting should be ripped out and replaced; this would be hopelessly expensive. The main aim should be to use modern-type, sensible lighting to replace equipment which has worn out and has to be replaced. The initial cost might be slightly higher, but the new lights are more economical, and would pay for themselves before long. At the moment it has been estimated that the wastage caused by upward-shining lights amounts to around £19,000,000 per year.

The guidance leaflets have been issued to all local councils, and some have taken it to heart. Consider The Avenue, in Southampton. Here the recommended lighting has been installed, with the result that the street itself is safely and evenly illuminated, with no glare, while the sky above remains black. If this example could be followed in every case, the British skies of, say, 2002 would be a great deal darker than they are in 1992.

Safety apart, what about the astronomical aspect? Spectroscopy is a major astronomical tool, and the spectral quality of outdoor lighting can play an important rôle in limiting interference. For example, there are some lights which spread across large regions of the electromagnetic spectrum. High-pressure mercury lighting is one example. On the other hand, low-pressure sodium lighting has its principal emission in a narrow line in the yellow region, and this can be tolerated, so that astronomers are quite happy to sacrifice a narrow part of the overall range. But things are much less satisfactory when we consider sports stadium lights, billboards, domestic indoor lighting leaking from uncurtained windows, and even car headlights! The University of London Observatory, at Mill Hill, has suffered badly; electronic devices can cope with some of the damage, but not all of it.

The pollution problem is not confined to visible light, and there are comparable problems in the radio range. Bands are reserved for the use of radio astronomers, and have been acknowledged by the International Telecommunications Union for many years, but it has come to the point where radio astronomers are badly restricted by having to confine themselves to their allotted bands; they need to observe right across the radio spectrum. Moreover, because of the rapidly developing radio

Light pollution! This photograph of the Earth at night shows the extent of the problem.

communications industry, there is strong pressure for new services such as mobile telephones and high-definition TV satellite communication to invade the protected wavebands. Orbital satellites are equally disruptive, notably the Glonass navigation vehicle which produces interference at one particular frequency of special importance (the line frequency of OH molecules).

Of course we all contribute to radio noise, simply by switching lights on and off, running motor-cycles, electric trains, microwave ovens, electric razors . . . the list is endless. Sir Bernard Lovell once told me that unless action is taken soon, radio astronomy from Earth will be a science limited entirely to the second half of the twentieth century. The only final solution appears to be to go to the far side of the Moon, which is always turned away from us and is therefore absolutely radio-quiet.

Next, what about space pollution? Military and communications satellites are already making their presence felt, but of even greater importance is the growing amount of space débris. Each satellite can stay aloft long after it has ceased to function. While it used to be true that 'what goes up, must come down', we are now putting more material into orbit than is wise, and much of it will remain there for centuries, thousands of years or even permanently.

Moreover, space-craft can collide, explode, and generate particles of a range of size, ranging from dust-flecks to large chunks of material as large as lorries. Because of the high relative velocity, a very small particle can inflict catastrophic damage. A piece of matter of mass

1 gramme, travelling at 18 miles per second, has an energy about the same as that of a 1-ton car travelling at 60 mph. Recently a Shuttle had to take evasive action to avoid a piece of space débris (actually, part of an old Russian Cosmos vehicle) which had fortunately been tracked, and another Shuttle had its window damaged by the impact of a flake of paint.

It is estimated that the Hubble Space Telescope has a 1 per cent chance of catastrophic damage by collision during its expected active lifetime of seventeen years, and the chances of minor damage are significantly greater. This is not acceptable, and as more and more débris accumulates in orbit the collision risks increase still further. In other words, low Earth-orbit is not now a good place in which to put delicate and expensive scientific instrumentation.

In addition, satellites (both active and dead) and large pieces of débris can be good reflectors of sunlight. They can be bright enough to cause trails on photographic plates, and many long-exposure plates are now ruined each year when they are found to be crawling with satellite tracks. Even worse, the sudden appearance of a bright satellite may ruin a very sensitive detector.

Worst of all are some commercial and military projects. Years ago, the infamous American 'Project West Ford' deliberately sent large numbers of copper needles into orbit so that they could act as an artificial ionosphere and help in long-range communication in time of war. Mercifully the needles came down—otherwise, they would have wrecked ground-based radio astronomy permanently—and the protests from the scientific community were so vehement that the experiment has not been repeated, but much more recently we have had to deal with two appalling suggestions: those of the Light Ring, and the Celestis Corporation.

The Light Ring, to mark the centenary of the Eiffel Tower, was to be a huge cross sent into space, tracking over the sky several times a night. The effects upon astronomical research can well be imagined. At about the same time, the Celestis Corporation of America planned to launch burnished cylinders carrying the remains of loved ones, so that they could be put into orbit as permanent reminders. Neither of these projects came to anything; French conceit is amazing, though for sheer bad taste the Celestis Corporation is in a class of its own. Let us hope that all future plans of this sort are outlawed. Otherwise, we might well have glaring satellites extolling the virtues of, say, patent medicines or cat-foods . . .

About natural brightness in the night sky we can do nothing, and neither would anyone wish to do so. There is the permanent airglow; we have the lovely polar lights or auroræ, as well as the ghostly glows of the Zodiacal Light and the Gegenschein—and, of course, the presence of the Moon makes dark-sky work impossible for parts of each month. But we must, surely, do all we can to limit the amount of artificial pollution. It is up to all of us to see that our children and our children's children can return to enjoying the beauties of a starlit sky.

 **30 THE GHOST PLANET**

Of all the programmes I have presented in the *Sky at Night* series since 1957, there are a few which stand out in my memory. One of these is certainly that of 8 December 1991, when I was joined by three radio astronomers headed by Professor Andrew Lyne, of Jodrell Bank. We discussed the discovery of the first known planet to have been detected in the system of another star, and it all seemed very exciting indeed. Later, it was established that there had been a major mistake, and that the planet did not exist; yet in the end, the whole episode reflected the greatest credit upon the astronomers who had been responsible for the error.

From my point of view the story began on 24 July 1991, when I was in Buenos Aires, in Argentina, at the General Assembly of the International Astronomical Union. The IAU meets once every three years, and this was its first foray into South America. For once I was wearing my 'astronomer's hat', as my presence in Argentina had nothing to do with television. I have been a full member of the Union for thirty years, and on this occasion I had been asked to edit the daily newspaper which is distributed every morning to each delegate. Normally, the editor has a team of assistants and translators, plus a full suite of offices. I had none of these, and to write an account of the whole saga would take a long time, particularly as in the end the Conference Centre caught fire!*—but at least I produced the paper, *Southern Cross*, on time every day.

When I arrived at the San Martín Centre early in the morning of 24 July, ready to start the day's work, I was greeted by Sir Francis Graham-Smith, the former Astronomer Royal. 'We have some news about the new planet!'

'What planet?' My thoughts flew instinctively to Planet X, the hypothetical world orbiting the Sun beyond the orbits of Neptune and Pluto, which I am sure exists. But this was not a planet in our Solar System. 'It's moving round a neutron star, 30,000 light-years away.'

I was taken aback. 'Where does this come from?'

'Jodrell Bank. Andrew Lyne and his team have found it with the Lovell Telescope'—the famous 250-foot 'dish'.

I asked only one more question. 'Have you got an article ready?'

'Yes. Here it is.'

'Right—you've got the front page!'

*See Chapter 58 of *Fireside Astronomy*, John Wiley & Sons, 1992.

The 250-foot Lovell radio telescope at Jodrell Bank.

To say that I was surprised would be an understatement. A neutron star is the last kind of place where one would expect to find a circling planet. I was also worried by the fact that the orbital period of the planet was given as six months, and I had an uneasy feeling that there might be some link with the Earth's own motion round the Sun. But I am no radio astronomer, and when we presented our programme in December the existence of the planet was regarded as well established.

A neutron star is a very strange object. According to accepted theory, it is the remnant of a supernova, i.e. a very massive star which has used up most of its fuel and has exploded, sending the bulk of its material way into space and leaving only a very small, super-dense core made up of neutrons. The average density is so great that a pin's-head of neutron star material would weigh as much as the *QE2*. Neutron stars spin rapidly, and send out pulses of radio emission from their magnetic poles; each time a beam sweeps over the Earth we receive a pulse—hence the term 'pulsar'.

A neutron star is believed to have a solid crust, lying over a fluid, probably super-fluid, interior. (The same is true of a raw egg. Spin it, stop it by putting your hand on it, and then let go; the egg will start to spin again of its own accord. Try it and see!) Studies of neutron stars are valuable in many ways. They are superb clocks; they have provided additional confirmation of relativity theory, and they are the best possible potential sources of gravity waves. It is therefore not surprising that they were very much on the Jodrell Bank programme, and in 1985 a concentrated search revealed forty new pulsars. Of these, 39 were conventional. The last one, known as PSR 1829-10 according to its co-ordinates in the sky (it actually lies in the little constellation of Scutum, the Shield) was not.

Normally, the pulses from a rotating neutron star can be timed very accurately indeed. The difference between the predicted and the observed time of a pulse is generally nil; in more scientific terms, there are no residuals. But this was not true of 1829-10, and a graduate student, Setnam Shemar, was asked to make some careful investigations. It was not long before he found something very interesting. Sometimes the pulses arrived earlier than had been predicted, while at other times they were late. Initially the residuals seemed to be random, but then, after some careful observations with the Lovell Telescope, Shemar found that

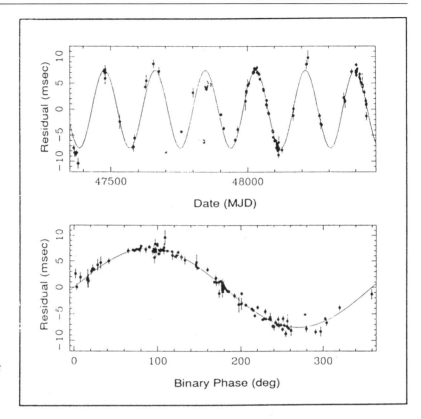

Graph taken to show the reality of the planet of the neutron star. Sadly, the planet does not exist.

they were cyclic. There was a definite 186-day period, and he was able to draw a curve of the residuals which was very different from the straight-line, nil-residual result from an ordinary pulsar.

Shemar decided that the cause of his phenomenon was movement. The neutron star, he surmised, was moving round the common centre of gravity of a binary system, and though the companion body could not be seen it was making its presence felt. When the neutron star was moving toward us (due allowance having been made for the overall motion in space) the pulses were early; six months later they were delayed, because the neutron star was then on the far side of its orbit and the pulses had further to travel. Obviously the observations were delicate, because the neutron star span three times per second, but the end conclusion was that 1829-10 was attended by a body about ten times as massive as the Earth. This is far too lightweight to be a star, and so it had to be classed as a planet. It was, Shemar said, moving in a circular orbit at a distance of around 67,000,000 miles from the neutron star, which is just about the same as the distance between our Sun and Venus.

All this was fascinating, but there were various obvious problems, and Andrew Lyne detailed three in particular. First, a giant star expands before exploding as a supernova, and this would be extremely dangerous for a planet moving in a close orbit; indeed, the luckless planet would presumably be engulfed, or at least evaporated. Secondly, the loss of mass after the supernova outburst would weaken the star's gravitational pull, and any orbiting body would be likely to fly off into space. And thirdly, there is the point that a supernova explosion is unbelievably violent. At its peak, the luminosity may be equal to that of a whole galaxy of stars—and a normal galaxy, such as ours, contains around 100,000 million suns. It is hard to see how any planet could survive such a holocaust.

Various efforts were made to find a solution. For example, it was suggested that the outburst which had led to the formation of this particular neutron star was much less than normal, so the dying star had presumably lost most of its mass at an earlier stage and there was not enough left to make a conventional supernova; instead, the stellar remnant subsided into a neutron star much more gently. Another idea was that the pulsar was not produced in the usual way, but was the result of a collision between two white dwarf stars which simply merged, providing so much mass that the nuclear material was forced into the neutron state.

But could the planet have been formed or captured after the outburst, not before? This was Lyne's favoured view during our television programme; if the material left over had formed into a thick disk, it could have produced the planet. There was also a chance that the planet was a newcomer to the system, and was simply captured from interstellar space, though this would involve some rather special circumstances.

I was bold enough to bring up my misgivings about the six-month period; could it be that the effects were due not to a planet, but to our own orbital motion round the Sun? If a cloud of gas lay in the same line of sight as the pulsar—possibly quite near the Solar System—signals from the pulsar would be delayed when they had to pass through the cloud, and timings would depend upon the exact alignment or non-alignment of the Earth. However, this was discounted, because the phase effects did not fit. All in all, it seemed at the time that a planet was the most plausible answer. Subsequently, papers were published speculating about the chances of life on this peculiar world!

It was only in January 1992 that the real explanation was found. One of the calculations had been made on the assumption that the Earth's orbit round the Sun is circular, but in fact it is not; the distance ranges between 91½ million miles in January out to 94½ million miles in July. When the calculations were re-worked, it became painfully clear that there was no need to assume the presence of any planet.

And this is where Andrew Lyne showed his absolute integrity. It would have been very easy to blame instrumental error or some sort of computer failure. He did nothing of the sort; he made an official statement explaining what had happened, and accepting full responsibility for it. A lesser man would have acted very differently, and I feel that the Jodrell Bank astronomers emerged from the whole episode with greater credit than they would have done if the bizarre planet had really existed instead of being a mere ghost.

31 THE MIGHTY KECK

O n Wednesday, 28 August 1991, I had my first sight of the Keck Telescope. It was breathtaking. I had seen most of the world's great telescopes, and used most of the main refractors during my Moon-mapping days, but the Keck was something new in my experience. It was so utterly unlike anything else.

The site itself is impressive enough: the top of Mauna Kea, the extinct Hawaiian volcano whose summit reaches to 13,796 feet above sea-level. (Since most of Mauna Kea lies below the ocean surface, the whole structure is considerably larger and more massive than Everest.) The Keck lies at 13,600 feet. From it the view is superb, because Mauna Kea is nothing if not majestic. In the distance you can see the crest of its twin, Mauna Loa, which is active at the present time; the lava from one of its associated volcanoes, Kilauea, often covers some of the main roads in Hawaii's Big Island. Indeed, an eruption was going on in the summer of 1991, and Chain of Craters Road, along which I had happily driven during my previous visit (in 1986) was blocked; the molten, red-hot lava pouring into the sea was most impressive. Years

Dome of the Keck Telescope on Mauna Kea, as I photographed it in 1991.

ago the lava even cut Saddle Road, which winds its way between the two huge volcanoes, and reached the outskirts of Hilo, the largest town on the island. Legend has it that disaster was averted only by the incantations of a powerful witch-doctor who had been hastily called in!

Extreme altitude has its disadvantages. The air is thin, so that one's lungs can take in only 39 per cent of the normal amount of oxygen, and this can be hazardous. Running around, or even moving quickly, is emphatically not to be recommended. Some people cannot tolerate the conditions at all. On this occasion I was with a BBC television team, and one visit to the summit was enough for the secretary who had come with us from London; she withdrew hastily, and retreated to the comfort of Hilo at sea-level.

Why, then, come to the top of Mauna Kea? The reason is that the thinness and (usually) dryness of the air is exactly what astronomers want. Atmosphere not only blocks and disturbs the light coming from space, but it also cuts out many of the most important wavelengths, notably a great deal of the infra-red. Atop Mauna Kea there is very little atmospheric water vapour left, which is why the site has been chosen for UKIRT, the United Kingdom Infra-red Telescope, as well as IRTF, an infra-red telescope facility operated by NASA. Various other telescopes have also been set up, such as the CFH (Canada–France–Hawaii) reflector. But so far, at least, the Keck is unique.

The world's largest really useful single-mirror telescope is the Palomar 200-inch, which was completed in 1948 and has contributed as much, if not more, to astronomy than any other telescope now in action. (The Russian 236-inch is, frankly, a failure; one Moscow astronomer recently said to me that in his view, it was 'a bad telescope on a bad site.') Refractors, which use lenses to collect their light, can never be as large as this, and the Yerkes Observatory 40-inch is never likely to be surpassed, because a larger lens supported round its edge would suffer unacceptable gravity strains. But there is also a limit for single mirrors for reflectors, partly because of atmospheric problems but also because of the sheer difficulty of making them. Very modern techniques, involving what is termed 'spin-casting', have improved things considerably, but there must be a limit. What can be done about it?

One tactic is to use several smaller mirrors, and combine them to produce a single image. The prototype here was the MMT or Multiple-Mirror Telescope at the Whipple Observatory, on Mount Hopkins in Arizona, where there were six 72-inch mirrors. The telescope was a success (the *Sky at Night* team paid a visit there in 1981) but the six mirrors are now to be replaced by a single, larger one. A second tactic is to have separate telescopes which can be used together; this is the plan for the VLT or Very Large Telescope which the European Southern Observatory is to set up in Chile. But there is another possibility, too. Make your mirror in segments.

Early experiments, mainly Italian, were carried out before the war, but were not successful; computers are essential, and there were no computers in the 1930s. But when the idea for a new telescope on Mauna

The Keck Telescope on Mauna Kea: When I took this photograph, in 1991, only 12 of the mirror's segments were in place.

Kea was mooted, one of America's most far-sighted astronomers, Dr Jerry Nelson, came to the conclusion that a segmented mirror was the answer. This was in 1977. Serious planning began in 1978, and the segmented pattern was approved in 1980. But what about funds?

This was where the W.M. Keck Foundation came into the picture. It is one of the largest American organizations which gives vast sums to charities or other good causes such as science, and in 1985 it donated 70 million dollars for the construction of the new telescope. This amounted to 75 per cent of the total cost. The rest was found, and the telescope went ahead. By August 1991, when I saw it, it was already operational, though incomplete. It looked like a typical artist's idea of a futuristic telescope, with masses of wires, ladders and platforms; but it had already shown that it worked well.

The diameter of the mirror is 10 metres or 33 feet. This is equivalent to 396 inches, almost twice the size of the Palomar mirror. To make such a giant out of a single piece of glass would be immensely difficult even with the spin technique, and so the Keck has 36 hexagonal segments which are put together to make up the correct curve and are aligned by a control system which has to be accurate to within a millionth of an inch—a thousand times thinner than a human hair. The whole telescope structure is 81 feet high, and together with the mirror weighs almost 300 tons. The collecting area is 818 square feet. Yet the total weight of glass involved is a mere 14.4 tons, as against 41 tons for the single mirror of the Russian 236-inch reflector. As for the Keck dome— well, it is 101 feet high and 122 feet wide, with a total moving weight of 700 tons. It is hardly necessary to add that the mounting is altazimuth; I fear that giant equatorials belong to the past. (I say 'I fear' because I am old-fashioned enough to be very attached to them!)

Because the mirror had to be made in sections, normal testing methods could not be used until the segments were suitably bent, which was done by stress techniques invented by Jerry Nelson specially for the purpose. Each segment is 6 feet in diameter and 3 inches thick, weighing 880 pounds; the material used is Zerodur, which is a low-expansion glass ceramic (that is to say, it does not expand and contract noticeably with changes in temperature, as most types of glass do). The focal length of the whole telescope is 57.4 feet.

Keeping the segments aligned, and to the correct shape, is of course the Keck's major problem, but early results were highly encouraging. 'First light' came on 24 November 1990, when only nine of the segments were in place; even so, the light-grasp was already equal to that of the Palomar 200-inch. By the time I arrived another three segments had been added, and another two were brought to the volcano-top while I was there. One advantage of the segmented pattern is that individual sections can be removed for re-polishing and adjusting without the need for putting the entire telescope out of action, as with a conventional instrument.

The control system depends on disk pads and what are called whiffletrees. A whiffletree (as if you didn't know!) is a vertical rod which branches out into twelve short posts; the name comes from a last-century pivoting crosspiece which allowed draft animals in a team to move independently while still pulling a cart. The whiffletrees and flex disks keep the mirror segments rigidly mounted against sideways motions while still allowing them to be moved for aligning the mirror array. In fact, the whiffletrees effectively 'float' the segments as if they were under zero gravity, absorbing strains which would otherwise deform them.

As I have said, the 36-segment array forms the correct optical curve, but each individual segment has a surface curvature which is not symmetrical. Jerry Nelson was quick to appreciate this, and so he invented a stressed-mirror polishing technique, in which the segments are bent by forces applied round their edges before being polished in the usual way. When the distorting forces are removed, the glass snaps back to the required non-symmetrical shape. (Simple—once you have thought it out!) When the glass has been polished, it is cut into hexagons; this makes it change shape slightly, but the effects are known, and can be allowed for. What are termed leaf springs then gently force the mirror into its final shape. Note also that the mirror cell, made up of over 1100 individual pieces, must hold its shape to within 1/25 of an inch, no matter where the telescope is pointing in the sky.

Finally, actuators in the form of screws produce the tiny adjustments needed to keep the segments aligned. Each segment has three actuators.

Whiffletrees!

The mirror control system runs continuously, and corrects the alignment of the array every half-second.

Quite clearly, the Keck is revolutionary in every sense of the term. 'First light', with the galaxy NGC 1232 as the target, was a great moment; NGC 1232 showed up magnificently across a distance of 65,000,000 light-years. Those who had cast doubt upon the feasibility of making a huge segmented mirror were proved wrong.

The tremendous light-grasp of the Keck means that it will be able to study incredibly faint objects, so remote that we see them as they used to be when the universe was young. It will also be able to obtain the spectra of very old stars in a way that no other telescope can match, and with the new infra-red techniques it will peer into those mysterious regions in which stars are being formed.

What else? The Keck should be able to search for planets around other stars. This was on the agenda for the Hubble Space Telescope, but had to be abandoned because of the faulty mirror. (Note, too, that the Hubble mirror is 'only' 94 inches in diameter, which is not much against the 396 inches of the Keck—though, of course, the HST has perfect seeing conditions all the time, whereas the Keck is still handicapped by the atmosphere remaining above Mauna Kea.) All in all, the possibilities of the Keck are tremendous.

There is another development, too. In April 1991 the W.M. Keck Foundation made a further grant of 75.4 million dollars for the construction of a twin telescope, which will be set up about 290 feet from the present dome. This means that Keck I and Keck II will be able to be used together, as an interferometer. When completed, the two telescopes in unison would be capable of distinguishing a car's headlights separately from a distance of 16,000 miles.

If the Keck twins come up to expectations—and there is no reason why they should not—they will represent one of the greatest triumphs of modern technology. I am glad to have seen Keck I at this early stage in its career—quite apart from the fact that there is no site in the world to rival Mauna Kea, where you have the feeling of being half-way into space.

32 TARGET JUPITER

It is now more than thirteen years since the Voyager probes encountered Jupiter, Voyager 1 on 5 March 1979 and Voyager 2 on the following 9 July. Both were completely successful, and of course both were only just starting their main careers. Since then two further probes to the Giant Planet have been launched, for very different reasons: Galileo on 18 October 1989, and Ulysses on 6 October 1990.

Galileo was a space-craft of new type, and was extremely ambitious. Unfortunately it could not go directly to Jupiter, as the Voyagers had done, because its launcher was not sufficiently powerful, and it was forced to follow a somewhat tortuous path. First it swung in toward the Sun, and on 7 February 1990 by-passed Venus at a distance of 1,700,000 miles, picking up speed by means of the now-familiar gravity-assist technique; while it was close to Venus, it sent back excellent pictures and data. It then moved outward, and passed by the Earth at 600 miles in December 1990. Its next foray was into the asteroid belt, and on 13 November 1991 it sent back the first close-range picture of one of these midget worlds; Asteroid 951 Gaspra proved to be wedge-shaped, not unlike a distorted potato, with a darkish, rocky surface pitted with craters. Galileo then came back to a further pass of the Earth, and

Notice board at JPL (the Jet Propulsion Laboratory), 3 February 1992, showing the current distances of the Galileo and Ulysses probes.

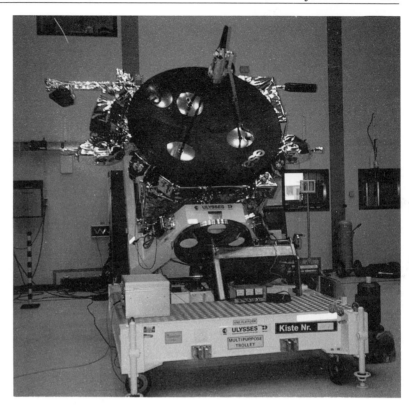

Ulysses: When I took this picture, I was standing within a few feet of the space-craft— which is now in orbit round the Sun.

if all goes well it will skim by us in December 1992 at a mere 200 miles before making its final outward trip to Jupiter, which it will reach in 1995. It is rather like driving from Brighton to Bognor Regis by way of Grimsby, but there was no alternative.

Galileo consists of two parts: an orbiter, scheduled to circle Jupiter for an extended period, and an entry probe, which will plunge into the outer clouds and send back data for the last minutes of its career before being destroyed. First, however, came the encounter with Venus. One picture, taken on 14 February 1990 though not transmitted until much later, was an improvement upon anything previously obtained. It showed that the distinct cloud layers which lie between 30 and 40 miles above Venus are in a state of considerable agitation, and also that the east–west banding of the clouds indicated high-speed winds of at least 230 m.p.h.

One of Galileo's instruments, NIMS (the Near Infra-red Mapping Spectrometer) was particularly useful for mapping Venus in this range of wavelengths, and demonstrated that there are layers of clouds, rich in sulphuric acid, at around 35 miles above the surface, i.e. several miles below the visible cloud-tops. In this way it is now possible to draw up a global portrait of the deep atmosphere of Venus. But so far as Galileo was concerned, Venus was incidental—and so for that matter was Gaspra. The main target was Jupiter.

The plan was quite clear-cut. Once in Jupiter orbit, the space-craft would begin a series of complicated manœuvres which would enable it to survey all the four large satellites, as well as some of the minor members of the Jovian family. Io, with its sulphur volcanoes, would be the orbiter's first target, and also its last. At the end of its scheduled career, Galileo would again plunge in to survey Io. This would mean entering the lethal radiation zones which encircle Jupiter, and it would be unlikely that the instruments would survive.

The entry probe would have a shorter but more spectacular career. It would enter the upper clouds at a height of about 60,000 miles above the planet, and would descend for about six hours, with peak deceleration at about 280 miles. At 55 miles the parachute drogue would open, and then the aft cover would be jettisoned to allow the parachute to open. At 30 miles the main aeroshell would separate, and the most important transmissions would begin; the probe would now have a life-expectancy of only a few minutes. The data would be relayed immediately via the orbiter, and would, it was hoped, go on for something like three-quarters of an hour, but by the time that the probe had descended to about 90 miles below the visible cloud-tops it would be certain to be crushed.

Unfortunately there was a serious hitch. The high-gain antenna, vital for sending back signals, failed to deploy properly, apparently because of insufficient lubrication. Frantic efforts were made to free it, but at the time when I write these words (August 1992) the antenna is still stuck. The low-gain antenna will work, of course, and all in all it should be possible to salvage most of the mission, but for the moment we can only hope for the best.

Ulysses, launched just under a year after Galileo, was of very different type. Though it encountered Jupiter in February 1992, it was not really a Jupiter probe at all. Its rôle was to survey the poles of the Sun.

Ulysses was given its final checks at Noordwijk in Holland, the headquarters of the European Space Agency, early in 1990. I saw it there, and stood within a few feet of it; it was sobering to realize that this space-craft would soon be in orbit round the Sun, millions of miles away from us, and would never return to Earth. Then, on 6 October of the same year, I was at Cape Canaveral watching the launch, from the Space Shuttle *Discovery*. Ulysses was on its way.

The solar poles are important, because conditions in those regions are different from those elsewhere, particularly with respect to the magnetic field. Near the Sun's equator everything is confused; high-speed particles overtake those moving at a slower rate, causing what may be termed a cosmic traffic-jam, and the magnetic field lines are convoluted. Near the poles, however, the situation was expected to be much more straightforward, and the magnetic field lines would be open, allowing solar wind particles to escape; here too it was thought that there would be 'coronal holes'—regions in the corona where the density

Antarctica: a mosaic of 40 images from Galileo, on 8 December 1990 (range 124,000 miles). The South Pole is to the left of centre.

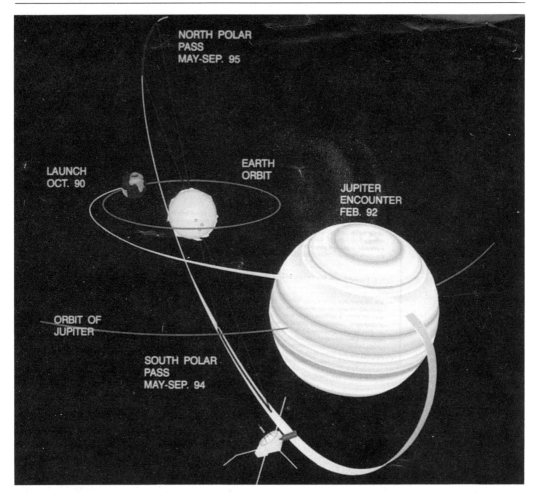

Path of Ulysses: the solar polar probe (European Space Agency).

was even lower than normal, so that solar wind particles could escape at a constant rate. Coronal holes had been discovered from the Skylab space-station as long ago as 1973, but our knowledge of them was (and still is) very incomplete.

Obviously we cannot study the Sun's poles from Earth, because we always see the solar globe more or less broadside-on, and no previous space-craft had been able to go very far either above or below the ecliptic in order to 'look down' or 'look up' at the Sun. This sort of manœuvre takes a great deal of planning, because it involves giving the probe very high velocity, and this in turn takes a tremendous amount of 'fuel' unless gravity-assist techniques are used. Moreover Ulysses, originally known as the OOE or Out of the Ecliptic Mission, had had a chequered history. It had been planned in 1974, but the launch vehicle was changed several times. The best idea, using a three-stage Centaur rocket, would have sent Ulysses direct to Jupiter, but after the *Challenger* disaster this had to be abandoned, and instead a less powerful launcher was substituted. Jupiter would not be by-passed until February 1992.

By direct injection into solar orbit it would not be possible to leave the ecliptic by more than 23 degrees, which was not enough. The only alternative was to use the pull of mighty Jupiter. So the scheme was to send Ulysses outward, make it pass over Jupiter's north pole, and then allow gravity to swing it southward, taking it to 70 degrees away from the ecliptic. It would then swing inward, passing over the Sun's south pole between June and November 1994 before crossing the ecliptic again in February 1995 and surveying the Sun's north pole between July and September of that year. Nominally the mission would then end, though the planners openly hoped that Ulysses would remain active for further polar passages nearer the end of the century, by which time the Sun would once more be near the peak of its eleven-year cycle.

The gravity-assist technique had been well tested, and there was no doubt that it was feasible insofar as Ulysses was concerned, but there were other problems to be taken into account. Jupiter is surrounded by strong zones of radiation, capable of killing any astronaut foolish enough to enter them. Of course there were no human travellers on Ulysses, but instruments are sensitive, as the planners well knew from previous experience. This was the main cause of the worry when I arrived at the Jet Propulsion Laboratory a few days before the Jupiter pass on 8 February. 'It's a stressful time,' said Peter Beech, the European Space Agency's operations manager, while Willis Meeks, the Ulysses project manager, commented that 'Jupiter's radiation is very hostile, so naturally we're concerned. You don't fly past Jupiter every day, and there are always unknowns to be concerned about'.

So far everything had gone well. Ulysses was on course; corrections to its trajectory had been made on 2 November 1990 and 8 July 1991, but a further adjustment in January 1992 had been found to be unnecessary. So we waited and hoped. Every possible precaution had been taken, and two of the most sensitive instruments had been switched off for the danger period, but nobody was over-confident.

There was also the problem of Io, Jupiter's innermost large satellite, which is very slightly bigger than our Moon. Io is a weird place. It has a red, sulphury surface, with violently active volcanoes which send out material—mainly sulphur and oxygen, with some sodium—at a rate of over 2000 pounds per second. The volcanoes had been detected by the Voyagers in 1979, and had presumably been active for an immense period. The oxygen and sulphur atoms are ionized by solar radiation, and form a dense plasma cloud along Io's orbit, known as a torus and less scientifically compared with a doughnut. It is 44,000 miles thick, and is probably the main source of the plasma associated with Jupiter. The Jovian magnetic field lines connect it to the rest of the Giant Planet's magnetosphere.

The danger period was due between 31 January and 16 February. Closest approach to Jupiter, say 235,000 miles above the cloud-tops (between the orbits of Io and the second large satellite, Europa) was timed for 4.01 hours Pacific Standard Time on 8 February, but the

The Moon from Galileo,
9 December 1990, from
350,000 miles: The
Mare Orientale is near
the centre; the Oceanus
Procellarum to the
upper right, with the
smaller Mare Humorum
below it. At the lower
left, along the southern
cratered highlands of the
lunar far side, is the
South Pole–Aitken
Basin.

worst moment would come five hours later, during the crossing of the Io torus.

Everything happened on schedule. Ulysses swept over Jupiter, and across the torus; at that time it was 416,458,914 miles from Earth and 501,978,560 miles from the Sun, with a velocity of 84,491 m.p.h. relative to the Earth and 61,249 m.p.h. relative to Jupiter. It had travelled a grand total of 624,000,000 miles since launch. The achievement was truly amazing, particularly since its initial departure from Earth had been faster than for any previous space-craft.

Though Ulysses was essentially a solar probe, the chance of carrying out new studies of Jupiter during the fly-by was much too good to be missed, and there was a great deal to learn. Ulysses carried no camera, and so no pictures were obtainable, but all the instruments on board worked perfectly. Unlike the previous Jupiter probes (two Pioneers and two Voyagers), Ulysses passed over the dusk section of the planet's magnetosphere, i.e. the region from which sunset would be seen, and the trajectory took it to high Jovian latitudes during its departure from the planet, so that it was possible to obtain unique data with regard to the Ionian torus.

Almost at once it was clear that new discoveries had in fact been made. The charged particles making up the radiation belts are constrained to follow the lines of force of Jupiter's magnetic field, so that the magnetometers on Ulysses could use the data to refine the mathematical models of the entire field. It was found that the outer magnetosphere does not rotate with the inner region; there is a sharp dividing line, so that the inner magnetosphere rotates with Jupiter while the outer part does not—an entirely unexpected result. All in all, the magnetosphere was even more complex than had been anticipated. There was another surprise, too, with the Io torus, which seemed to be less uniform than it had been during the Voyager missions; it was incomplete, and broken up into patches, possibly indicating that the Ionian volcanoes were less active than they had been in 1979 (though this is not to suggest that they are dying out; no doubt the changes are periodical, as is the case with the volcanoes of Earth).

Within a few hours of the crossing of the torus the main danger was regarded as over, and at a party that evening, which I attended, there was what Willis Meeks called 'a lot of handshaking, a lot of smiles and a lot of jubilation', while Edward Stone, the director of JPL commented that 'this was an historic moment for Ulysses'. The poles of the Sun lay ahead.

There were, of course, other investigations to be carried out before then. Between 27 February and 18 March Ulysses and the Sun would be on opposite sides of the Earth, and this was the time to search for gravity waves, which had long been theoretical possibilities but which had never been detected. A gravity wave striking Ulysses would alter its position by about one centimetre. Amazingly, this would cause a slight shift in frequency of the radio signals, according to the well-known Doppler effect, which could be detected at JPL. There would also be studies of cosmic ray particles entering the Solar System from outer space, which would be expected to be more easily detected coming in from the directions of the Sun's poles than when coming in from the direction of the solar equator. Dust particles spread around the Solar System would also be under scrutiny. Only after that would the main part of the mission begin.

It is rather surprising to find that Ulysses will not go close to the Sun. Indeed, it will never again be as near as it was when I stood beside it at Noordwijk. It is the angle of view which matters, and even during the next crossing of the ecliptic, in February 1995, the distance between Ulysses and the Sun will still be 200,000,000 miles.

Gaspra, from the Galileo space-craft: the first close-range picture of an asteroid (13 November 1991, from a range of 9900 miles).

It will never come home; after it ends its active career it will remain in orbit, endlessly circling the Sun and possibly lasting for as long as the Earth itself. We have no means of knowing its eventual fate. But as I left JPL it was clear that the outlook was good. Ulysses was another historic 'first', and we all look forward to exciting new data from it over the coming years.

33 AMONG THE ASTEROIDS

M uch has been heard recently of the small 'Earth-grazing' asteroids, which skim past us and sometimes cause alarm in the popular Press. But we must not forget the more normal asteroids, those of the main belt, which are admittedly junior members of the Sun's family, but which are far from lacking in interest. By now over 4000 of them have had their orbits worked out, though only one (Ceres) is as much as 500 miles in diameter. Only six have diameters of more than 200 miles:

1	Ceres	584 miles
4	Vesta	358 miles
2	Pallas	327 miles (mean value; Pallas is triaxial)
10	Hygeia	267 miles
511	Davida	239 miles
704	Interamnia	210 miles

There has been considerable discussion about the origin of the asteroids. Originally it was believed that they were the débris of a larger planet or planets which had broken up, but today this idea is rather out of favour. For one thing, there is no common point in the orbits where the break-up would have occurred; secondly, all the asteroids put together would not make up one body as massive as the Moon; and thirdly, the different types of asteroids indicate that there would have had to have been several different parent bodies. Therefore, it seems more logical to assume that no large planet could form in that region of the Solar System because of the powerful disruptive influence of Jupiter. If a sizeable planet started to form, Jupiter promptly broke it up, so that the end product was a swarm of dwarfs.

By now we have found out a good deal about the composition of the asteroids, and we can learn more from studies of meteorites; it seems very probable that asteroids and meteorites are identical—for example, the object which hit Arizona over 20,000 years ago could have been classed either as a large meteorite or as a small asteroid. Spectroscopic studies tell us that there are at least three or four different types of asteroids. Those of Type S are basically stony, not unlike the mantles of the terrestrial planets and of the Moon. Type M asteroids are primarily metallic; those of Type C are carbonaceous, and those of Type D are dark, their colour possibly coming from complex hydrocarbon molecules

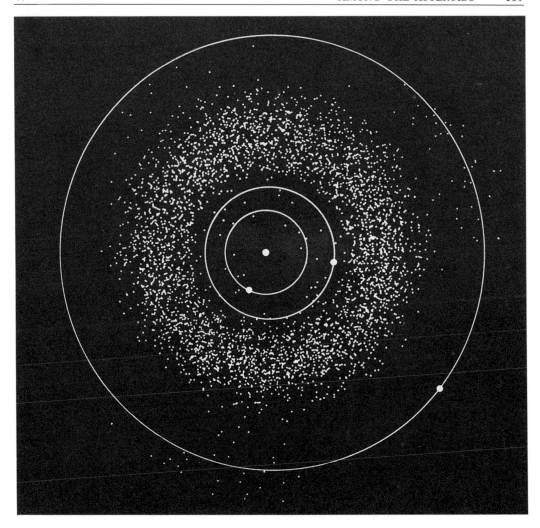

Distribution of the asteroids. Most, but by no means all, are in the main zone between the paths of Mars and Jupiter. The orbits shown are those of the Earth, Mars and Jupiter.

on their surfaces. In general, asteroids in the further part of the main belt are redder than those which are closer to the Sun.

There are several well-defined asteroid families, the members of which may well have been produced by the break-up of a parent body. Gaps in the zone were predicted by Daniel Kirkwood in 1857, and have been fully confirmed; again they are due to the perturbing influence of Jupiter. The main families are those of Hungaria (distance 1.9 astronomical units from the Sun),★ Flora (2.2 units), Koronis (2.9 units), Eos (3.0 units) and Themis (3.1 units); further out we find the Hilda family (4.0 units) and the isolated asteroid 279 Thule (4.2 units or about 395,000,000 miles), which has a D-type spectrum and is about 80 miles in diameter. Thule has a revolution period of 8.2 years, and may well be the brightest member of another family.

★As a reminder; one astronomical unit is equal to the mean distance between the Earth and the Sun, in round figures 93,000,000 miles or 150,000,000 kilometres.

A typical asteroid: 3477 Kazbegi shown as a white line near the centre of the picture. 30-minute exposure with the 1-metre Schmidt telescope at La Silla (European Southern Observatory). Kazbegi is about 5 miles across; magnitude 17.5; period 3 years 7 months.

Even further out there are more than a hundred 'Trojan' asteroids, which move in the same orbit as Jupiter, keeping either 60 degrees ahead of the Giant Planet or 60 degrees behind it, so that they are in no danger of being swallowed up. The Trojans provide some interesting dynamics. If one of them moves slightly closer-in to the Sun than Jupiter, it will move faster, and so will catch Jupiter up. However, since Jupiter will then be ahead of the asteroid its powerful pull will tug the smaller body forward, so that its orbital speed will increase. But increasing speed means making the orbit larger than Jupiter's, so that it will move more slowly and will lag behind again. The result is that a Trojan will oscillate quite safely around a mean point. Some of the planetary satellites have Trojan-like companions; thus in Saturn's system the relatively massive satellite has one Trojan (Helene), and Tethys two (Telesto and Calypso). Mars has one known Trojan asteroid, 1990 MB.

Efforts have been made to find an Earth Trojan, but with no success. However, the asteroid Toro has an orbit not too unlike that of the Earth, and when it was discovered there were even reports that it was a second Earth satellite. This is quite wrong. Toro is in resonance with the Earth, which means that the Earth/Sun/Toro configuration repeats itself at regular intervals, but Toro is a perfectly normal asteroid, and its orbit is decidedly more eccentric than ours.

A real oddity is 2060 Chiron, which has a period of 50 years and spends almost all its time between the orbits of Saturn and Uranus. Its diameter is uncertain, but it is probably large by asteroidal standards, and recently it has been found to develop what seems to be a temporary atmosphere. Its distance from the Sun ranges between 794,000,000 miles and 1,757,000,000 miles, and its next perihelion passage is due in 1996, so that it is receiving almost as much heat from the Sun as it ever can,

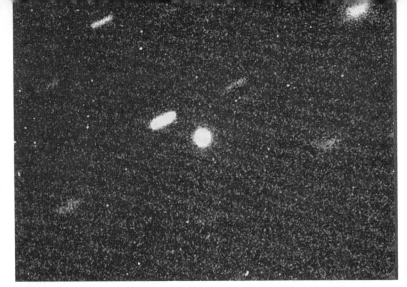

Asteroid 5145 Pholus (originally 1992 AD): Image with the Danish telescope at La Silla: O. Hainaut and A. Smette; reproduced by kind permission of the European Southern Observatory. The telescope followed the asteroid, whose magnitude was 16.7; the image is round, while the stars are shown as short trails. The diffuse trail at the upper right corner is a galaxy. The picture was taken on 5 February 1992.

and presumably some of the icy material on its surface has started to evaporate. It is much too large to be a giant comet, as has been suggested, and probably it is better regarded as a surviving planetesimal, i.e. a fragment of the original material from which the main planets were built up. It is not impossible that Pluto and its satellite Charon may be of the same basic type as Chiron, though they are considerably larger. (Do not confuse Chiron with Charon; it is a pity that the two names are so alike.) Another curious object is asteroid 5145, Pholus, whose path takes it out well beyond Neptune, and which has a revolution period of 93 years. It may be as large as Chiron, and is presumably of the same nature.

Yet another curious object was found in August 1992 by David Jewitt and Jane Luu, using the 2.2-metre telescope on Mauna Kea. They detected a dim speck of light—magnitude 23—which proved to be very remote, moving round the Sun at a distance of around 3700 million miles, slightly further out than the mean distance of Pluto and well beyond Neptune. The period is probably about 260 years. As yet we know little about it, but it seems to be reddish and asteroidal, with a diameter of the order of 125 to 200 miles. It is starting to look as though there may be a whole crop of 'planetesimals' in the far reaches of the Solar System. The newcomer will certainly be named in the near future; its provisional designation is 1992 QB1.

There have been various proposals for sending missions to asteroids, and even mining them. This may be possible in the future, but certainly not yet. Undoubtedly an asteroid would be a strange place to visit; from the main belt there would be many neighbours visible with the naked eye, but the astronaut would have practically no weight, because the gravitational pull would be so low, and a space-craft encounter with an asteroid would be more in the nature of a docking operation than a conventional landing.

If you have a telescope, a set of tables and a star map, it is interesting to go asteroid-hunting. You will be surprised how many of them you can pick up on a clear night!

34 THIRTY-FIFTH ANNIVERSARY

The first *Sky at Night* programme, broadcast in April 1957, was ushered in by the spiked comet, Arend-Roland, which was a very prominent object precisely at that time. I do not remember whether I regarded it as a good-luck sign, but at any rate the programme 'caught on', and we celebrated our thirty-fifth anniversary in April 1992. I did my best to sum up some of the amazing triumphs which have taken place over this period. Clearly I can do no more than skim the surface, and many people will disagree with my choices, but it is entertaining to look back and see the main events of each year during the lifetime of the *Sky at Night*. So here is my personal selection:

1957. Start of the Space Age, on 4 October. The Russians launched Sputnik 1, the pioneer artificial satellite, which carried very little apart from a transmitter but which was of unsurpassed importance. It showed, once and for all, that we can break free from the Earth. It also acted as a stimulus; at that time the American embryo space programme was frankly floundering, and tension increased with the ascent of a much heavier sputnik later in the year.

1958. Launch of the first US satellite. Explorer 1. Compared with the Sputniks it was a pygmy, but it was responsible for the first major discovery of the Space Age: the identification of the Van Allen radiation belts which encircle the Earth.

1959. The first 'Moon year', again Russian dominated. Luna 1 flew past the Moon in January, confirming (among other things) that the Moon has no overall magnetic field. In September Luna 2 crash-landed on the Moon, and in October Luna 3 went on a round trip, sending back the first pictures of the Moon's far side, which we can never see from Earth because it is always turned away from us. I was in fact the first to show these pictures in Britain; they were sent through to me from Moscow when I was actually 'on the air'. I was frankly proud to recall that the Russians had used my maps of the lunar 'edge' to correlate the Luna 3 images with the side of the Moon which we have always known.

1960. A quieter year in space, but it did see the ascent of the first meteorological satellite, Tiros 1. On the ground, Martin Ryle and Anthony Hewish, the great radio astronomers, developed the method of aperture synthesis which has been of such vital importance in research.

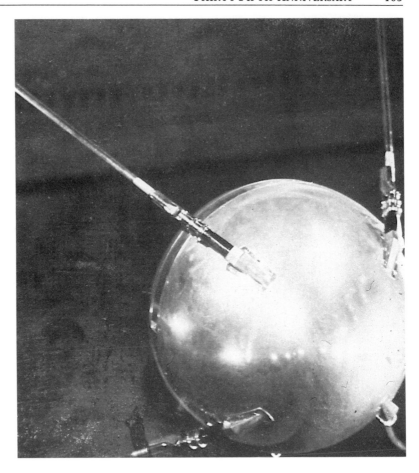

Sputnik 1: The first artificial satellite; launched on 4 October 1957, ushering in the Space Age.

1961. The first man in space; in April Yuri Gagarin, in Vostok 1, completed a full orbit, at once disposing of several of the much-discussed 'bogeys' (space sickness, giddiness, cosmic ray and meteoroid damage, etc.). The Russians also launched the first interplanetary probe, Venera 1, though they soon lost touch with it, and its final fate is unknown. In Australia, the 210-foot radio telescope at Parkes was completed.

1962. An eventful year. John Glenn became the first American to orbit the Earth; Mariner 2 made a pass of Venus, and sent back the first close-range information from that curious world; Telstar became the first television satellite to link Europe with America, and instruments carried aloft in a rocket identified the first cosmic X-ray source, Scorpius X-1.

1963. The first year in which two manned space-craft were in orbit simultaneously; both were Russian, one piloted by Andriyan Nikolayev and the other by Pavel Popovich. Valentina Tereshkova became the first woman in space. Astronomically, the main event was the identification of the first quasar, 3C-273, by Maarten Schmidt at Palomar.

Earthrise! The crescent Earth appears over the Moon (Apollo 14).

1964. At last the American lunar programme achieved a major success; Ranger 7 sent back excellent close-range pictures before crashing on to the surface. All the previous Rangers had failed for one reason or another.

1965. The first close-range pictures were obtained from Mars, by Mariner 4, showing that the Red Planet is cratered—and finally disposing of the canal myth. Alexei Leonov made the first 'space-walk'. However, the most important event of the year was certainly the identification of the cosmic background radiation, by Penzias and Wilson in the United States.

1966. A 'Solar System' year. Luna 9 came down gently on to the Moon, showing that the surface is quite firm enough to bear the weight of a space-craft; Orbiter 1 obtained the first really detailed lunar pictures, and the Russians landed Venera 3 on Venus, though they failed to keep in touch with it until it hit the surface.

1967. The Russians continued their investigations of Venus, and Venera 4 sent back signals during its descent through the planet's atmosphere, though once more contact was lost before touch-down. At Cambridge, the first pulsar was identified by the radio astronomer Jocelyn Bell-Burnell.

1968. Apollo 7, the first manned mission of the series. Astronauts Schirra, Eisele and Cunningham successfully tested it in Earth orbit.

On the Moon: The Lunar Roving Vehicle of Apollo 15 at the foothills of the Lunar Apennines.

Radio astronomers in Australia discovered the Vela pulsar, the remnant of a supernova which must have blazed out not too long before historic times.

1969. The Apollo Year. In March No. 9 was tested in Earth orbit; in May No. 10 took Stafford, Young and Cernan round the Moon, and then, on 20 July, Apollo 11 deposited Neil Armstrong and Edwin Aldrin in the Mare Tranquillitatis. The barrier between the two worlds had at last been broken down. Apollo 12 followed in November, and Astronauts Conrad and Bean were able to bring back pieces of an old Surveyor soft-landing probe which had been on the Moon for some years. The main astronomical event was the optical identification of the Crab pulsar by astronomers at the Steward Observatory in America.

1970. The near-tragedy of Apollo 13; on the outward journey there was a major explosion in the service module, and it was only by a combination of courage, skill and luck that the astronauts returned safely. In December came the launch of the first satellite to be devoted entirely to X-ray work, Uhuru.

1971. Two more Apollos: No. 14 (January) and No. 15 (July). The astronauts of Apollo 15 (Scott and Irwin) were the first to drive around in an LRV or Lunar Roving Vehicle. Mariner 9 encountered Mars in

November, and was put into a closed orbit; over the next months it sent back a series of spectacular pictures, giving us our first views of the Martian volcanoes.

1972. End of the Apollo programme, with Nos. 16 (April) and 17 (December); Harrison Schmitt, on Apollo 17, became the first professional scientist to go to the Moon.

1973. The first close-range pictures from Jupiter sent back by Pioneer 10. The Skylab space-station was launched, and its first crews occupied it.

1974. The first two-planet probe, Mariner 10, which encountered first Venus and then Mercury; the Mercurian craters were confirmed. Pioneer 11 obtained further images from Jupiter. In Australia, the Anglo-Australian Telescope was completed and set up at Siding Spring, New South Wales.

1975. The first pictures received direct from the surface of Venus, by the Russian vehicle Venera 9.

1976. The first controlled landings on Mars, with Vikings 1 and 2. The main objective was to search for life, but no positive signs of activity were found. The Russian 236-inch reflector was completed, though it has never been a success.

1977. A very eventful year. The Voyagers were launched toward the outer planets; the rings of Uranus were discovered by the occultation method, and Charles Kowal discovered the curious 'asteroid' Chiron. From Siding Spring, the Vela pulsar was identified with an excessively faint optical object.

1978. Launch of the IUE (International Ultra-violet Explorer) satellite, which operated until well into the 1990s even though its official life-expectancy was a mere three years. The X-ray satellite known as the Einstein Observatory was also sent up, and two Pioneer probes were launched toward Venus.

1979. News from the outer planets. Pioneer 11 made a preliminary pass of Saturn, while first Voyager 1 and then Voyager 2 encountered Jupiter, sending back pictures and data which far surpassed anything received from the Pioneers. The Los Muchachos Observatory on La Palma was officially opened. From Mauna Kea Pluto and Charon were imaged separately; this was the first actual proof of Charon's existence.

1980. Voyager 1 flew past Saturn, obtaining the first detailed close-range information. It also studied Titan, finding that the cloudy, nitrogen-methane atmosphere hides the surface completely.

1981. Voyager 2 encountered Saturn, confirming and augmenting the Voyager 1 information. It did not make a close approach to Titan, because it was programmed to go on to the outer giants Uranus and Neptune.

1982. Successful soft landings on Venus by Veneras 13 and 14; pictures sent back. From Palomar, Halley's Comet was recovered, and was therefore seen for the first time since 1911.

Facing page: The surface of Mercury: a mosaic of photographs taken by Pioneer 10.

1983. The 'IRAS Year'; launch and active career of the immensely successful Infra-Red Astronomical Satellite. Identification of the first millisecond pulsar, PKS 1937+215, which spins 642 times per second.

1984. The SMM (Solar Maximum Mission) satellite repaired in space, after having been captured and taken into the Shuttle. The Isaac Newton Telescope installed at La Palma.

1985. The two Russian Vega probes to Halley's Comet encountered Venus, dropping balloons into the atmosphere of the planet. The first 'Einstein ring' discovered (a gravitational lens effect).

1986. A year of mixed fortunes. The Halley armada reached its target; all five vehicles (two Japanese, two Russian and one European) were successful. Voyager 2 encountered Uranus, and performed faultlessly; but there was also the *Challenger* disaster, when the Shuttle blew up and killed its crew. Quite apart from the human tragedy (which was of course much more important than anything else), the accident put the progress of American space research back by several years.

The frozen lake of Triton: This is a computer-generated picture, from Voyager 2 pictures, and shows what would actually be seen by an observer flying over the 'lake'.

1987. The William Herschel Telescope installed at La Palma. Outburst of SN 1987A in the Large Cloud of Magellan, the first naked-eye supernova to have been recorded since Kepler's Star of 1604.

1988. Completion of the longest space mission to date; V. Titov and M. Manorov came down after spending 366 days on the Mir space-station. The Australia Telescope (a network of linked radio telescopes) was completed—but the famous radio telescope at Green Bank, West

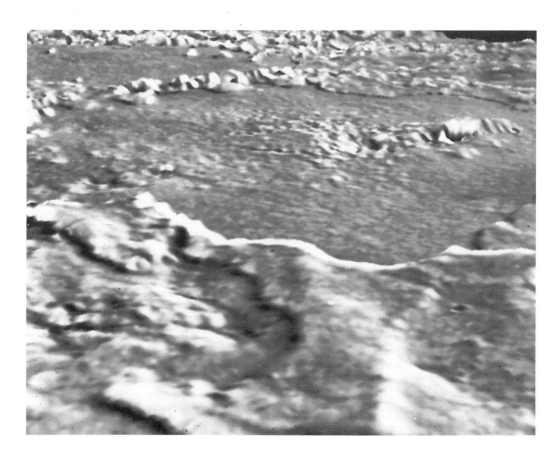

Virginia, collapsed without warning, fortunately without causing any casualties.

1989. Voyager 2 encountered Neptune, sending back superb pictures and data from Neptune itself, Triton and the other members of the system. The Hipparcos astrometric satellite was launched; it was put into the wrong orbit, but skilful manœuvring meant that the entire programme could be carried out. Two interplanetary probes were also launched, Magellan to Venus and Galileo to Jupiter.

1990. The Hubble Space Telescope launched (25 April), and subsequently found to have a faulty mirror. The Rosat (Röntgen) X-ray and EUV satellite was launched. First light was obtained with the Keck telescope on Mauna Kea. A bright white spot appeared on Saturn; and at La Palma, astronomers recorded the first surface details on a star (Betelgeux).

1991. The Compton gamma-ray observatory satellite was launched, while Rosat began a survey of the sky at EUV (Extreme Ultra-violet) wavelengths. Spectacular views of Venus were obtained by the orbiting Magellan probe, and Galileo obtained the first close-range picture of an asteroid (Gaspra).

1992. The Ulysses space-craft flew past Jupiter, using gravity assist techniques to throw it into an orbit well out of the ecliptic so that it could survey the poles of the Sun. The Giotto vehicle, which had passed through the head of Halley's Comet in 1986, encountered a second comet (Grigg-Skjellerup). COBE, the Cosmic Background Explorer satellite, detected slight variations in the microwave background radiation which were of tremendous significance from a cosmological point of view.

It is an impressive catalogue of achievement. By the year 2000 we may well have permanent space-stations, and perhaps even a base on the Moon. And looking back still further, it is amazing to realize how quickly all this has happened. The first airman, Orville Wright, made his initial 'hop' in 1903; the first space-man, Yuri Gagarin, went up in 1961, and the first man on the Moon, Neil Armstrong, landed there in 1969. I know (or knew) all three, so that in a way I rather feel that I span the ages. I wonder what will happen during the next thirty-five years? I am afraid that I will not be presenting *The Sky at Night* in 2027, unless of course I live to the advanced age of a hundred and four, but I am quite sure that there will be plenty to say.

From my little observatory in Sussex I have watched what has been going on, and I have tried to share it. I hope I have succeeded—and I thank all those who have followed the programmes throughout the years.

35 THE BACKGROUND OF THE UNIVERSE

At the present time there are not many astronomical discoveries which make headline news in the tabloid Press. One which did so was that of April 1992, in a rather unexpected way. It arose from results sent back by a very special satellite: COBE, the Cosmic Background Explorer, which was launched in 1989 into a near-polar orbit which takes it round the Earth at a height of 560 miles. Its main task was to study what is called the microwave background, made up of weak radiation coming in from all directions all the time, and believed to be the very last manifestation of the event in which the universe was created, some time between 15,000 million and 20,000 million years ago.

This event was the 'Big Bang', in which everything—space, time, matter; everything there is—came suddenly into existence. Because space was created at the same moment, we cannot say where the Big Bang happened. It happened 'everywhere'. Neither can we ask what happened before it, because there was no 'before'. What we have to do is to try to trace the story through to the present epoch. We cannot go back further than 10^{-45} of a second before the event, because all our laws of science break down, but we can do our best to follow things through from there.

When the universe was born it was not only small, but was intensely hot. Immediately afterwards there was a period of very rapid inflation, when the universe not only expanded but also cooled down. Initially there was a mish-mash of particles, but about five minutes after the Big Bang we come to the creation of atomic nuclei out of the primordial protons and neutrons. Unattached neutrons decay with a half-life of ten minutes or so, but neutrons locked up in atomic nuclei do not. Heavier nuclei were formed. It took half a million years for the temperature to fall sufficiently to allow electrons to stick to the atomic nuclei and form true atoms. Before that, radiation could not escape; it was blocked. When atoms were formed, however, the radiation was free. This was the time of the so-called de-coupling, when the temperature of the universe had dropped to a mere 2.7 degrees above absolute zero (absolute zero being -273 degrees Centigrade, the coldest temperature there can possibly be).

So far as we are concerned, the story continues with the work carried out in 1965 by two American radio astronomers, Arno Penzias and Robert Wilson. They were using a special kind of radio telescope (a horn antenna) for some new research when they found, to their surprise, that

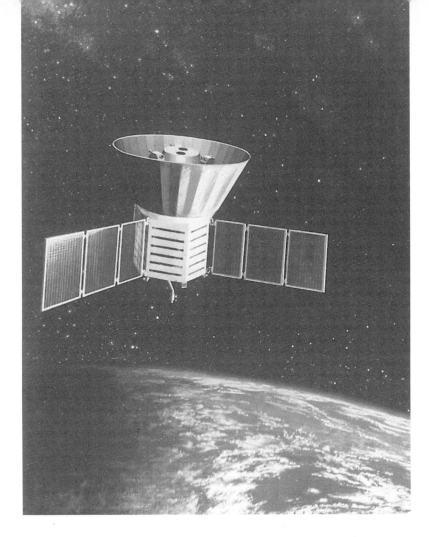

COBE (the Cosmic Background Explorer).

they were detecting a steady 'hiss' in the microwave region of the electromagnetic spectrum. For some time they could not identify it, but eventually they realized that it represented the last traceable effects of the Big Bang. Moreover, it had been predicted on theoretical grounds by R.H. Dicke, who had even been preparing to search for it when he heard of the work by Penzias and Wilson.

This was satisfactory enough, but there was an immediate problem. So far as could be ascertained, the background radiation was completely uniform; it arrived at exactly the same intensity from all directions. This indicated a completely 'smooth' universe, and it followed that during the early inflationary period the universe must also have been smooth. Yet how could our current, obviously non-smooth universe develop from such a uniform start? It did not seem to make sense.

Astronomers did their best to trace irregularities in the microwave background. Work carried out at Jodrell Bank, and by Cambridge workers associated with the observatory on Tenerife, seemed initially promising, but it was then found that the slight irregularities they believed they had detected were due simply to cosmic rays in the halo of our own Galaxy.

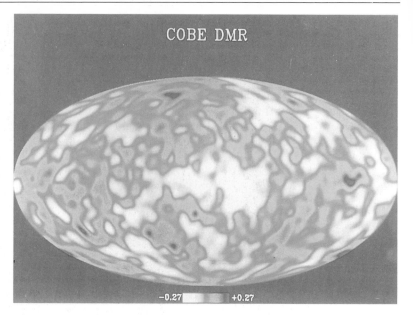

COBE DMR

−0.27   +0.27

All-sky map from COBE (the Cosmic Background Explorer), 1992.

This is why COBE was launched. It carried three main instruments, two of which were cooled down to a temperature of below 2 degrees above absolute zero so that their sensors should not be blinded by their own heat emission. By careful design it was hoped to produce temperature maps of the sky accurate to 0.0003 of a degree—far better than anything achieved before. Between 1989 and the spring of 1992 COBE made over 300,000,000 measurements, and, to the general relief, variations were found. They were very slight, which is why they had escaped detection earlier—they amounted to about ten parts in a million—but at least they were there.

This made things much easier to understand. Variations in temperature also meant variations in density. COBE had found evidence of 'ripples', or rarefied wisps of material, which were described as being the largest and most ancient structures in the universe, extending in some cases to 60 billion trillion miles. With this uneven distribution of density, gravitational effects could come into play, so that clumps of matter could draw together to produce groups of galaxies, then galaxies, stars, planets and—ultimately—you and me. Moreover, COBE was recording phenomena dating back to a mere 300,000 years after the Big Bang itself.

Astronomers were enthusiastic. John Mather, one of the principal investigators, said that 'what we have found solves a major mystery, revealing for the first time the primæval seeds which developed into the modern universe.' Stephen Hawking described it as 'the discovery of the century, if not of all time'; Jasper Wall commented that 'this is a wonderful discovery for us. It is a confirmation of a picture of the hot Big Bang universe to which we have subscribed for some time.' (Typically, senior members of the Church were less ecstatic. The Bishop of Peterborough said that 'this doesn't make a great deal of difference

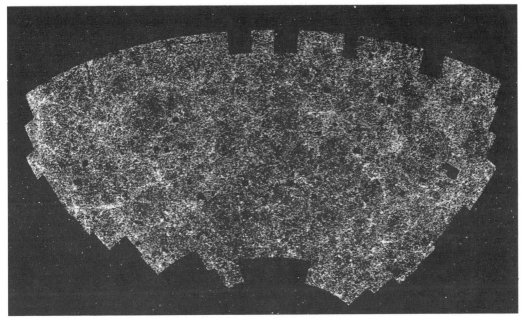

The distribution of 2 million galaxies shown as a map made up of many tiny dots. Each dot represents a small patch of sky containing many images. Dots are black where there are no galaxies, white where there are more than 20, and grey for a number between one and 19. The overall mottled pattern is caused by small-scale clustering of the galaxies. The small bright patches are individual galaxy clusters. The larger elongated bright areas are super-clusters and filaments. These surround darker voids where there are fewer galaxies than average. Photo: Department of Astrophysics, University of Oxford.

to me. It certainly doesn't make any difference to God.' I cannot follow the logic behind this, but no doubt I have missed something!)

There were other links, too. It has long been thought that there is a great deal of invisible matter in the universe, made up of material which is beyond our normal experience and which we cannot see—though efforts are being made to detect it, notably at Boulby Mine near Whitby, where a research station has been established in a potash/salt mine 3000 feet below ground level to screen the equipment from cosmic rays. COBE has helped here, too.

Can we be absolutely certain about these interpretations? A note of caution has been sounded by the Astronomer Royal, who points out that the universe is a dusty place; we see dust in nebulæ, between the stars, and even in the Solar System. The variations detected by COBE are so tiny that dust could be involved. However, at the moment it is fair to say that the results look extremely promising.

If we are right about the ripples detected by COBE, we have a much clearer understanding of what happened after the Big Bang, but we are no wiser about the Big Bang itself. We have no idea how it happened, or why. Whether we will ever find out in the future remains to be seen; for now, we have to admit that while we are strong on detail; we remain woefully weak on fundamentals.

At least COBE has given us new leads. Even if it does no more, it has a secure place in scientific history, and it stresses, once again, the vital links between ground-based astronomy and space research.

36 COMETARY ENCOUNTERS OF THE SECOND KIND

History was made on 10 July 1992, when the gallant space-craft Giotto encountered the periodical comet Grigg–Skjellerup—more than six years after its dangerous rendezvous with Comet Halley. Nobody expected Giotto to be able to perform adequately; in the event it surpassed all hopes. Apart from the camera, which was blind because part of its baffle had fallen off and was blocking its lens, the experiments worked well.

To recapitulate: Giotto was launched from Kourou, in French Guyana, on 2 July 1985. Its task was to go right into the head of Halley's Comet, and send back the first close-range pictures of a cometary nucleus—in fact the first clear pictures of any kind, because when a comet nears the Sun, and is close enough to be studied in detail from Earth, it hides its nucleus behind a cloudy veil due to the evaporating 'ices' and dust particles. Also en route to Halley were two Japanese probes (Suisei and Sakigake) and two Russians (Vega 1 and Vega 2). All four space-craft were successful. Giotto, built in Bristol by British Aerospace, went right through the coma, and on the night of March 13–14 1986 passed within 376 miles of the nucleus, which proved to be shaped rather like an avocado pear and measured $9.3 \times 5 \times 5$ miles. The total volume was over 500 cubic kilometres, with a mass of between 50,000 and 100,000 million tons (it would take 60,000,000,000 Halley's Comets to equal the mass of the Earth). Unfortunately, a particle about the size of a grain of rice hit Giotto some fourteen seconds before closest approach, when the space-craft was still just over a thousand miles from the nucleus. Though communication was soon re-established, the camera never worked again—to the intense disappointment of its Principal Investigator, Dr Horst-Uwe Keller.

Halley is very large for a short-period comet. Therefore it is somewhat atypical, and astronomers badly wanted to obtain data from a different sort of comet. Grigg–Skjellerup was the ideal candidate. First, however, let me say something about comets in general.

They are the most erratic members of the Solar System. They are of low mass, and the only substantial part of a comet is its nucleus, which has been described as 'a dirty ice-ball'. Only when a comet nears

Comet Grigg–Skjellerup, imaged on 29 June 1992 with the NTT (New Technology Telescope) at the La Silla Observatory, Chile. Reproduced by permission of the European Southern Observatory.

the Sun does it develop a head, and sometimes a tail or tails; there may be an ion tail pushed out by the pressure of sunlight, and/or a dust tail, repelled by the solar wind. According to the Dutch astronomer Jan Oort, comets are very ancient objects which have been left over, so to speak, from the very earliest chapter in the story of the Solar System. They lose mass every time they pass through perihelion (Halley's Comet loses 300,000,000 tons at each return), and unless the supply of short-period comets were constantly replenished there would soon be none left. Oort supposes that there is a vast storehouse of comets moving round the Sun at a distance of a light-year or so. When a comet in the Oort cloud is perturbed for some reason or other (perhaps by a passing star, or, in my view more probably, a very remote planet), it starts to swing inward toward the Sun, one of several things may then happen to it. It may simply pass by and return to the Oort cloud; it may fall into the Sun and be destroyed, as happened to Comet Howard-Kooman-Michels

Comet Grigg–Skjellerup, imaged on 10 July 1992 with the NTT (New Technology Telescope) at the La Silla Observatory, Chile. Reproduced by permissionof the European Southern Observatory.

in 1979; it may be perturbed by a planet, usually Jupiter, and thrown out of the Solar System, as with Comet Arend–Roland of 1957 and the brilliant Comet West of 1976; or it may be forced into a much shorter-period orbit, so that it will return regularly. More than 120 short-period comets are now known, most of which have periods of less than 30 years with their aphelia at about the same distance as the path of Jupiter. Only Halley ever becomes brilliant, though it signally failed to do so in 1986 and will be no better at the next return, that of 2061. The average lifetime of a short-period comet moving in the inner Solar System is probably from about 10,000 years to 100,000 years.

Because of their low mass, comets are very easily 'pulled around' by planets, and their orbits may be drastically altered. Lexell's Comet was bright in 1770, when it made a close approach to the Earth, but a subsequent encounter with Jupiter changed its path to such an extent that it has never been seen again.

There are other factors, too. The movements of comets are affected by the jets of dust and gas escaping from the nucleus, producing a sort of 'rocket effect' as predicted by Fred Whipple in the 1950s. Encke's Comet, which has a period of only 3.3 years and has been seen at over fifty returns (in fact it can now be followed all round its orbit), is a case in point. From its discovery in 1786 until about 1825 the period decreased by 2½ hours at each revolution, but today the decrease is only a few minutes per revolution, possibly because the comet's nucleus is very irregular in shape and is wobbling as it spins, like a top.

The ageing of a short-period comet is not a straightforward process. The surface of the nucleus is covered by a dark crust or mantle of dust, and only a small fraction of this surface is active at any one time, as was demonstrated during the Halley pass. Where there are cracks in the crust, the underlying dirty ice is exposed to the Sun and the ices sublimate, producing the gas and dust jets. As well as the steady erosion and loss of the surface layers as the comet grows older, bits of the dust mantle may be blown off by the gas-jets, producing surges in activity. Sudden outbursts in the brightness of comets by a factor of 100 or more have been observed—as with Comet Tuttle–Giacobini–Kresák, which in May and July 1973 twice brightened 10,000-fold. And we all remember the 300-fold flare of Halley's Comet in February 1991.

Biela's Comet is the most famous case of a break-up. It divided in 1845 and has not been seen since 1852, though its débris produced brilliant meteor showers in 1772 and again in 1885 (the shower is more or less defunct now). Pairs of known comets may be the fragments of a larger body (Neujmin 3/van Biesbroeck, and 1987 Levy/1988 Shoemaker–Holt). The Kreutz sun-grazing comets probably had a giant progenitor, and two of them, the Great Comet of 1882 and Ikeya–Seki of 1965, were observed to split, while West's Comet of 1976 had a nucleus which broke up into five parts in March of that year. Some comets were lost because their orbits are insufficiently well-known; the classic example is Swift–Tuttle, the parent comet of the Perseid meteor stream, which was seen in 1862 and had a computed period of 120 years,

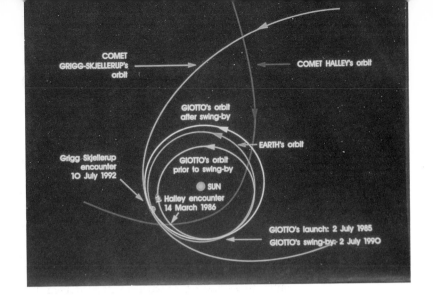

COMET GRIGG-SKJELLERUP's orbit

COMET HALLEY's orbit

GIOTTO's orbit after swing-by

EARTH's orbit

Grigg Skjellerup encounter 10 July 1992

GIOTTO's orbit prior to swing-by

SUN

Halley encounter 14 March 1986

GIOTTO's launch: 2 July 1985
GIOTTO's swing-by: 2 July 1990

Path of Giotto: encounters with two comets—Halley and Grigg-Skjellerup.

but failed to appear in 1982. It finally turned up in 1992, so that the period was 130 years instead of the expected 120.

Grigg-Skjellerup was discovered on 23 July 1902 by the New Zealand amateur John Grigg. The magnitude was then 9.5, and it was followed until 3 August. On 17 May 1922 J.F. Skjellerup, an Australian living in South Africa, discovered a 12th-magnitude comet which proved to be identical with Grigg's, and since then it has been seen at every return, though it has never become brighter than magnitude 9. It could be identical with a comet seen by J.L. Pons as long ago as 1808.

Grigg-Skjellerup is very different from Halley. It is an old, worn-out comet which seldom produces anything in the way of a tail, though its mile-wide nucleus is associated with a dust-cloud of elliptical form. The comet was expected to be far less active than Halley's, which was one reason why astronomers were so interested in studying it.

Now let us return to Giotto.

Everyone was delighted that the probe survived the Halley encounter, and plans were made for GEM, the Giotto Extended Mission. Various comets were considered. Two were serious candidates: DuToit-Hartley and Hartley 2, but of these the first would be too far from the Sun at the time of encounter, and the second moved in a path which was not known with sufficient precision. So Grigg-Skjellerup was elected.

The first step was to put Giotto into hibernation. Everything was closed down by 2 April 1986, and the space-craft was left to its own

The Giotto probe, half-built; I took this photograph of it at Bristol, on 19 January 1984, where the space-craft was being constructed. I am wearing a 'clean suit', as the whole area had to be kept completely dust-free.

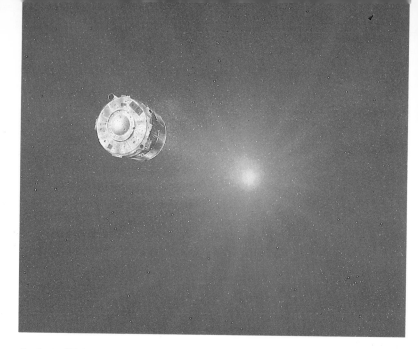

Cometary encounter: Impression of Giotto passing Comet Grigg-Skjellerup (Paul Doherty).

Notice at Darmstadt, before the Giotto pass of Grigg-Skjellerup. At that time Giotto was still 'in hibernation'.

devices. Nobody had any idea whether it could ever be 'woken up'; the planners simply had to hope for the best. Re-activation was initiated on 19 February 1990 and to the relief of all concerned Giotto responded; by 25 February it was announced that the re-activation was complete apart from the few inoperative instruments, notably the camera. On 2 July Giotto flew past the Earth at a distance of 14,127 miles, and the Earth's gravity was used to put it into a path which would take it to Grigg-Skjellerup. Again it was put into hibernation, and woken up for the second time on 4 May 1992, when it was 136,000,000 miles from the Earth—considerably further away from us than the Sun.

Tension mounted at Darmstadt, headquarters of the mission, as Giotto closed in. On the evening of 10 July the closest approach took place, and again everything which could work, did. Susan McKenna-Lawlor reported exciting results from her energetic-particle experiment, and showed that Grigg-Skjellerup was even more unlike Halley than had been expected. There was a good deal of gas, plus fine dust, and the bow-shock was very marked. Giotto suffered four major hits, one of them from a particle about half a millimetre in diameter, but this time there was no damage.

Obviously the results will take time to analyse, but there will be plenty of information available, and all in all this second encounter has been almost as valuable as the first. Is the story over yet? Possibly not. If there is enough fuel reserve, then Giotto may be sent on to a third comet in a few years' time. And in any case, its 13½-month orbit will automatically bring it back to the neighbourhood of the Earth every 108 years. It is conceivable that in the future some space scientist will locate it and literally fish it down. At the moment this is beyond our powers, but it may not always be so, and I would like to think that Giotto could be rescued.

For the moment we must wait. But Giotto has plenty of time.

INDEX